# Field Mathematics for Electromagnetics, Photonics, and Materials Science

## A Guide for the Scientist and Engineer

## Tutorial Texts Series

- *Metrics for High-Quality Specular Surfaces*, Lionel R. Baker, TT65
- *Field Mathematics for Electromagnetics, Photonics, and Materials Science*, Bernard Maxum, TT64
- *High-Fidelity Medical Imaging Displays*, Aldo Badano, Michael J. Flynn, and Jerzy Kanicki, TT63
- *Diffractive Optics–Design, Fabrication, and Test,* Donald C. O'Shea, Thomas J. Suleski, Alan D. Kathman, and Dennis W. Prather, TT62
- *Fourier-Transform Spectroscopy Instrumentation Engineering,* Vidi Saptari, TT61
- *The Power- and Energy-Handling Capability of Optical Materials, Components, and Systems,* Roger M. Wood, TT60
- *Hands-on Morphological Image Processing,* Edward R. Dougherty, Roberto A. Lotufo, TT59
- *Integrated Optomechanical Analysis,* Keith B. Doyle, Victor L. Genberg, Gregory J. Michels, Vol. TT58
- *Thin-Film Design: Modulated Thickness and Other Stopband Design Methods,* Bruce Perilloux, Vol. TT57
- *Optische Grundlagen für Infrarotsysteme,* Max J. Riedl, Vol. TT56
- *An Engineering Introduction to Biotechnology,* J. Patrick Fitch, Vol. TT55
- *Image Performance in CRT Displays,* Kenneth Compton, Vol. TT54
- *Introduction to Laser Diode-Pumped Solid State Lasers,* Richard Scheps, Vol. TT53
- *Modulation Transfer Function in Optical and Electro-Optical Systems*, Glenn D. Boreman, Vol. TT52
- *Uncooled Thermal Imaging Arrays, Systems, and Applications*, Paul W. Kruse, Vol. TT51
- *Fundamentals of Antennas*, Christos G. Christodoulou and Parveen Wahid, Vol. TT50
- *Basics of Spectroscopy*, David W. Ball, Vol. TT49
- *Optical Design Fundamentals for Infrared Systems, Second Edition,* Max J. Riedl, Vol. TT48
- *Resolution Enhancement Techniques in Optical Lithography*, Alfred Kwok-Kit Wong, Vol. TT47
- *Copper Interconnect Technology*, Christoph Steinbrüchel and Barry L. Chin, Vol. TT46
- *Optical Design for Visual Systems*, Bruce H. Walker, Vol. TT45
- *Fundamentals of Contamination Control*, Alan C. Tribble, Vol. TT44
- *Evolutionary Computation: Principles and Practice for Signal Processing*, David Fogel, Vol. TT43
- *Infrared Optics and Zoom Lenses,* Allen Mann, Vol. TT42
- *Introduction to Adaptive Optics,* Robert K. Tyson, Vol. TT41
- *Fractal and Wavelet Image Compression Techniques,* Stephen Welstead, Vol. TT40
- *Analysis of Sampled Imaging Systems,* R. H. Vollmerhausen and R. G. Driggers, Vol. TT39
- *Tissue Optics: Light Scattering Methods and Instruments for Medical Diagnosis*, Valery Tuchin, Vol. TT38
- *Fundamentos de Electro-Óptica para Ingenieros,* Glenn D. Boreman, translated by Javier Alda, Vol. TT37
- *Infrared Design Examples,* William L. Wolfe, Vol. TT36
- *Sensor and Data Fusion Concepts and Applications, Second Edition*, L. A. Klein, Vol. TT35
- *Practical Applications of Infrared Thermal Sensing and Imaging Equipment, Second Edition*, Herbert Kaplan, Vol. TT34
- *Fundamentals of Machine Vision*, Harley R. Myler, Vol. TT33
- *Design and Mounting of Prisms and Small Mirrors in Optical Instruments*, Paul R. Yoder, Jr., Vol. TT32
- *Basic Electro-Optics for Electrical Engineers,* Glenn D. Boreman, Vol. TT31
- *Optical Engineering Fundamentals,* Bruce H. Walker, Vol. TT30
- *Introduction to Radiometry,* William L. Wolfe, Vol. TT29
- *Lithography Process Control*, Harry J. Levinson, Vol. TT28
- *An Introduction to Interpretation of Graphic Images,* Sergey Ablameyko, Vol. TT27
- *Thermal Infrared Characterization of Ground Targets and Backgrounds,* P. Jacobs, Vol. TT26
- *Introduction to Imaging Spectrometers,* William L. Wolfe, Vol. TT25
- *Introduction to Infrared System Design,* William L. Wolfe, Vol. TT24
- *Introduction to Computer-based Imaging Systems,* D. Sinha, E. R. Dougherty, Vol. TT23
- *Optical Communication Receiver Design,* Stephen B. Alexander, Vol. TT22
- *Mounting Lenses in Optical Instruments,* Paul R. Yoder, Jr., Vol. TT21

# Field Mathematics for Electromagnetics, Photonics, and Materials Science

## A Guide for the Scientist and Engineer

Bernard Maxum

Tutorial Texts in Optical Engineering
Volume TT64

SPIE
PRESS

Bellingham, Washington USA

Library of Congress Cataloging-in-Publication Data

Maxum, Bernard.
Field mathematics for electromagnetics, photonics, and materials science : a guide for the scientist and engineer / Bernard Maxum.
p. cm. – (Tutorial texts in optical engineering ; TT64)
Includes bibliographical references and index.
ISBN 0-8194-5523-7 (softcover)
1. Engineering mathematics. 2. Vector analysis. I. Title. II. Series.

TA330.M38 2004
620'.001'51563—dc22

2004016206

Published by

SPIE—The International Society for Optical Engineering
P.O. Box 10
Bellingham, Washington 98227-0010 USA
Phone: +1 360 676 3290
Fax: +1 360 647 1445
Email: spie@spie.org
Web: http://spie.org



Printed in the United States of America.
Fifth printing

The International Society
for Optical Engineering

to my wife

**Marilyn Jo**

# Introduction to the Series

Since its conception in 1989, the Tutorial Texts series has grown to more than 60 titles covering many diverse fields of science and engineering. When the series was started, the goal of the series was to provide a way to make the material presented in SPIE short courses available to those who could not attend, and to provide a reference text for those who could. Many of the texts in this series are generated from notes that were presented during these short courses. But as stand-alone documents, short course notes do not generally serve the student or reader well. Short course notes typically are developed on the assumption that supporting material will be presented verbally to complement the notes, which are generally written in summary form to highlight key technical topics and therefore are not intended as stand-alone documents. Additionally, the figures, tables, and other graphically formatted information accompanying the notes require the further explanation given during the instructor's lecture. Thus, by adding the appropriate detail presented during the lecture, the course material can be read and used independently in a tutorial fashion.

What separates the books in this series from other technical monographs and textbooks is the way in which the material is presented. To keep in line with the tutorial nature of the series, many of the topics presented in these texts are followed by detailed examples that further explain the concepts presented. Many pictures and illustrations are included with each text and, where appropriate, tabular reference data are also included.

The topics within the series have grown from the initial areas of geometrical optics, optical detectors, and image processing to include the emerging fields of nanotechnology, biomedical optics, and micromachining. When a proposal for a text is received, each proposal is evaluated to determine the relevance of the proposed topic. This initial reviewing process has been very helpful to authors in identifying, early in the writing process, the need for additional material or other changes in approach that would serve to strengthen the text. Once a manuscript is completed, it is peer reviewed to ensure that chapters communicate accurately the essential ingredients of the processes and technologies under discussion.

It is my goal to maintain the style and quality of books in the series, and to further expand the topic areas to include new emerging fields as they become of interest to our reading audience.

*Arthur R. Weeks, Jr.*
*University of Central Florida*

# Table of Contents

## Chapter 2 Vector Algebra Review / 2-1



## Chapter 3 Elementary Tensor Analysis / 3-1

# List of Figures

## Chapter 1



## Chapter 2



## Chapter 4

## Chapter 5



## Appendix B

# List of Examples and Applications

Throughout this book, several examples and applications illustrate the use of the described concepts in engineering and the physical sciences. These are by no means comprehensive, but are provided to illustrate the utility of vector and tensor ideas.

# Acknowledgments

The author is indebted to Drs. Vladislov A. Bevc and Andrey Beyle who reviewed the entire manuscript and provided many useful ideas and suggestions especially regarding the mathematics. Consultation with Dr. Alex Matheson at an early stage in the book's formulation and several useful suggestions by Vito Florimonte are gratefully acknowledged. Jane Stanley and Debby Piper typed the first draft, and Crystal Basham did the equations and graphics for the original manuscript. Students Ganesh Ramanth, Emir Suljkanovic, and Pramod Varma assisted with the final figures. The author is most grateful and appreciative to each of these fine people. Finally, I am deeply indebted to my loving and patient wife who provided consistent encouragement and critique throughout the preparation of this book and who gave up many evenings and weekends that we might otherwise have spent together.

Bernard Maxum
October 2004

# Preface

The overriding objective of this book is to offer a review of vector calculus needed for the physical sciences and engineering. This review includes necessary excursions into tensor analysis intended as the reader's first exposure to tensors, making aspects of them understandable at the undergraduate level. A secondary objective of this book is to prepare the reader for more advanced studies in these areas.

As the world embarks on new horizons in photonics and materials science, honing one's skills in vector calculus and learning the essential role that tensors play are paramount. New inroads in engineering are driving the need for a revamp of engineering mathematics in these areas. Profound new paradigms in optical engineering and new advances in composites are necessitating these changes. The author has found that there is an ever-increasing need for vector calculus concepts to be extended to tensors and that his undergraduates can indeed grasp tensorial concepts if taught following the lines of thinking presented here.

Whereas the classical approach to teaching electromagnetics at the junior level has been to avoid any mention of tensors, the high-tech world entering the third millennium warrants a rethinking of this practice. This is especially true as nonlinear optical effects become more common in the design of optical systems. Advanced materials, especially composites and nanodesigned materials, provide further evidence supporting the teaching of tensor fundamentals to upper-division* students. Even for isotropic materials, the fundamental relationship between stress, strain, and elastic modulus—which are rank-two and rank-four tensors—requires a fundamental understanding of tensor analysis. For anisotropic materials such as composites, piezoelectric materials, and magnetostrictive materials, tensorial relationships are unavoidable even in the linear regime.

Furthermore, the development of new photonics devices in optoelectronics, acousto-optics, magneto-optics, and fiber optics is playing an ever-increasing role in contemporary communications system design.[1,2,3,4] Pollock states

---

* University-level juniors and seniors.

*The drive for faster systems has led to... [an] electronic speed bottleneck...This has motivated the study of integrated optics, where light, which has a much higher implicit frequency limit, is used to control light... Without a doubt the biggest research task...will be the development of optical switches and devices, and better communication architectures.*

These devices include laser sources,[5] optical switches, rare-earth-doped fiber amplifiers,[6] nonlinear-effect fiber amplifiers,[7] nonlinear-effect fiber soliton waves,[8] optical detectors,[9] and new dispersion-managed optical fibers.

## Uses of this Guide

This is a guide, and was not planned as a text book. As such, it is intended for multiple uses, including its use as a

1. reference to salient differential and integral forms for problem solving,
2. supplement to an engineering or science course, used in conjunction with and as a counterpart to it,
3. study guide before entering such courses,
4. reference manual in an R&D laboratory or design group,
5. complement to required or elective math courses, or just as a
6. refresher and reference source to vector calculus and an introduction to tensor analysis, or a
7. text, provided the instructor devises problem sets to provide the usual practical experience with numerical examples.

## Who is this guide written for?

Many students and working professionals experience a new awakening when they see and feel first-hand how complex mathematical concepts are applied to understanding real-world challenges. It is the intent of this guide to provide some of the mathematical prowess to facilitate reaching this level of professional elation. Other ways to state this are

*Mathematics is fun!*

or

*Knowledge is power!*

Courses such as electromagnetics (commonly called "fields") are often viewed by students as tough and something to be avoided until late in their program. Such postponement is not advised. Other courses, such as quantum physics, fiber optic communications, nonlinear fiber optics, structural analysis, materials science or any of a number of other engineering and physical science courses are understood through exposure to the concepts of vector and tensor calculus. It is hoped that this type of exposure will provide the confidence needed to encourage students to complete mathematically intensive courses earlier in their programs by allaying their fear of an imagined mathematical abyss. In this way they will be better prepared for more advanced studies.

John R. Whinnery in his classic paper[10] "The Teaching of Electromagnetics" states

> *The set of four equations we know as "Maxwell's equations," in modern notation, is simple enough to imprint on a T-shirt, and yet rich enough to provide new insights throughout a lifetime of study. Some students grasp the clarity, power and excitement on first introduction while others have a very rough time with the concepts.*

Whinnery's paper is intended to give students encouragement in approaching electromagnetics with clarity and excitement and to seek its power. His remarks might also be applied in varying degrees to other areas of physics as well, especially with regard to the mathematical constructs of Schrödinger, which are necessary for understanding the quantum physics inherent in the optical devices cited above as well as in nonlinear optical constructs.

Other challenging areas contributing to new millennium technology include

- optical communications,
- homeland security sensor systems,
- optical materials design,
- new applications of bi-anisotropic materials,
- optically based computer design for ultra-high speed and data throughput,
- space-based materials development,
- new innovations in medical imaging.,
- the design of ultra-high-bandwidth ultra-dense multi-access networks and their associated components, and
- crystal physics.

These are but few of many that could be listed.

This guide is also for anyone who is, or endeavors to be, involved in research, development, or education relating to these and other new frontiers in science and engineering. Although this guide is written with explanations and examples intended for the upper-division and first-year graduate student in science or engineering, it is also intended for those engaged in graduate research and in industrial research and development who have already been exposed to some of the concepts.

While excursions into tensors were originally written with undergraduates in mind, the author has discovered that many professionals, including academics, have a restricted understanding of tensors. A glimpse of the tensor-dyadic issue in the introduction to Chapter 3 and the rank-order issue in Section 3.1 (including the footnote), a study of Table 3-1 (at the end of Chapter 3), Table 4-1 (in Section 4.6), and especially Table 4-2 in Section 4.7 may open doors for some and provide good instructional fodder for anyone who uses tensors in their upper-division or introductory graduate courses.

## Content

This guide consists of five chapters and four appendices. As an introduction, Chapter 1 deals with a suggested notation that distinguishes between scalars, vectors, phasors, dyadics, and higher rank tensors, without the use of boldface characters. In so doing, it briefly covers other typical notational forms that are used in this book or that one may encounter in the literature. It also covers spatial differentials and the concept, definition, and use of partial derivatives. This includes the general formulation of partial derivatives of unit vectors with respect to coordinates—a factor often neglected in undergraduate instruction leading to incorrect answers. A simple example of this concept is provided.

Chapter 2 provides a review of vector algebra covering variant and invariant scalars, scalar and vector fields, the notation and utility of phasor scalars, phasor vectors, phasor dyadics, and phasor tensors in general. Classical arithmetic vector operations of addition, subtraction, and dot-, cross-, and direct-product operations are discussed along with physical applications of these. Open and closed line and surface integrals of vector fields are cited as being potent uses of dot products in integral calculus covered in Chapter 5. Vector field direction lines and equivalue surfaces of scalar fields are also developed as further examples of the power of cross- and dot-product operations. In the process, the need for metric coefficients in coordinate expansions is introduced.

Chapter 3 gives an introduction to tensors, and the power of the use of tensor analysis is explained at a level intended for the junior, senior, or early graduate student, who may not have been previously exposed to dyadics or other tensors beyond scalars and vectors. The concept of *inner product*—a term used synonymously with dot product—is discussed. The dot products of a dyadic with a vector and a vector with a dyadic are carried out in detail, and in the process the adjective "inner" is made apparent. The dot and double-dot products of two dyadics are also detailed. These inner-product operations are expressed in their considerably more simplified tensor notation in order to illustrate the value and power of the latter.

The chapter introduces tensors of higher rank (through examples in the mechanics of materials and nonlinear optics) and the interpretation of rank in terms of "*directional compoundedness*"—a term coined by the author to help those unfamiliar with tensors to overcome the idea that a quantity can have more than one direction at every point in space and time. The rudiments of tensor analysis include rules for term-by-term rank consistency and rules for determining the resulting rank after performing certain product operations. This concept is detailed and tabulated in Chapter 4.

Chapter 4 is a review of vector calculus differential forms with excursions into tensor analysis. First-order vector differential operators are introduced with a historical perspective on the use of the "del" operator. Scalar differential operators, differential equations, and eigenvalues are generally discussed. The concepts of gradient, divergence, and curl are described in physical terms and developed from their basic definitions without the use of coordinate systems. The rank of the resultants of these first-order vector differential operators is tabulated in Table 4-1.

Vector operators of vector operators, such as the Laplacian of scalar and vector fields and six others that are commonly used in junior-level courses, are also explained in terms that conjure up images of the fields and the effects of these operators on those fields. These second-order operations are tabulated in Table 4-2 and developed in generalized orthogonal curvilinear coordinates. These are then reduced to cylindrical coordinates (rather than the usual rush to Cartesian) in order to illustrate certain terms that otherwise disappear when Cartesian coordinates are used—cylindrical being the simplest of the non-Cartesian systems, and also coincidentally being the most appropriate in the analyses of optical fibers.

Chapter 5 deals with integral forms of vector calculus and also with excursions into tensor calculus. It first delineates line and surface integrals of scalar, vector, and tensor fields with dot-, cross-, and direct-product integrands. It then covers Gauss' divergence theorem and Stokes' curl theorem with examples of their applications. These are first explained in physical terms and then developed mathematically. Four of the most common forms of Green's identities are then presented, and Green's functions are offered as a powerful approach to solving inhomogeneous partial differential equations.

## Appendices

A. This appendix serves as a supplement to the vector arithmetics* covered in Section 2.4. The commutative and associative laws of vector addition and subtraction cited in Section 2.4 are demonstrated. As an application, these laws are used to show graphically and mathematically how vectors may be bisected. (Other applications of vector arithmetics can be found in Chapter 2.)

B. In this appendix divergence and curl are developed from their definitions in the more conventional Cartesian coordinates for further clarity of the concepts covered in Sections 4.4 and 4.5. The divergence is developed again in cylindrical coordinates as a first-level generalization towards curvilinear coordinates taking into account that the azimuthal $\phi$ coordinate is the sole curvilinear coordinate in the cylindrical system. Coordinate conversions and differentials, metric coefficients, differential elements of length, and equations of coordinate surfaces are tabulated for various orthogonal coordinate systems. Finally, graphical representations of the coordinate surfaces for each specific coordinate system are displayed in perspective view following each table.

C. Intermediate-level tensor calculus is used in this appendix for the purpose of demonstrating several issues and rules cited in Chapter 3 and for providing proofs of several important postulations used in Chapter 4, especially in Tables 4.1 and 4.2. At this level we intended it for those who have learned the concepts in the earlier chapters or for those already familiar with the area. These include the proof of the Lagrange identity [Eq. (4.7-15)] that is often presented to upper-division students without such a proof. The appendix also demonstrates that the divergence operator cited by Eq. (4.7-7) is not only valid when applied to vector and dyadic

* Pronounced arith·**met′**ics

operands [given by Eqs. (4.4-22) and (4.7-9), respectively] but also to any tensor of general rank. Finally, to offset the tendency to treat the divergence, curl, and gradient as analogous to the dot, cross, and direct products, we emphasize that there are two properties of the nabla vector differential operator that must both be taken into account. That is, for all but Cartesian coordinates, the analogy is false.

D. Appendix D provides Cartesian and cylindrical coordinate expansions of first- and second-order vector differential operators acting on scalar (where appropriate), vector, and dyadic operands. Two applications from materials science are presented that require the taking of the curl of the dyadic strain and the gradient of the dyadic stress. The first yields another dyadic, which in turn is an application of the dyadic line integral Eq. (5.1-4). The second yields a 27-term triadic, which is explicitly provided in Cartesian coordinates [Eq. (D.1-10)] and cylindrical coordinates [Eq. (D.2-10)]. Several of the more common Cartesian and cylindrical coordinate expansions presented in this appendix are listed on the inside front and back covers of this book for the readers' convenience.

## Glossary

A glossary of the acronyms, terms, and definitions used in this book precedes the index.

## References

1. Paul E. Green, Jr., *Fiber Optic Networks*, Prentice-Hall, Englewood Cliffs, NJ (1993).
2. Stuart Personik, *Fiber Optics Technology and Applications*, Plenum Press, New York (1985).
3. Chin-Lin Chen, *Elements of Optoelectronics and Fiber Optics*, Richard D. Irwin, Inc., Burr Ridge, IL (1996).
4. Clifford R. Pollock, *Fundamentals of Optoelectronics*, Richard D. Irwin, Inc., Burr Ridge, IL (1995).
5. *ibid*, p. 435ff.
6. Chen, *op. cit.*, pp. 153-155, 564ff.
7. Pollock, *op. cit.*, p. 195ff.
8. Pollock, *op. cit.*, p. 204ff.
9. Such as avalanche photodiodes, Pollock, *op. cit.*, p. 379ff.
10. John R. Whinnery, "The teaching of electromagnetics," *IEEE Trans on. Ed.*, **33**(1) pp. 3-7 (1990).

# Chapter 1
# Introduction

As an introduction to this guide, three topics are briefly reviewed. First, a convenient, consistent, and pedagogically functional notation is provided and various other notational approaches that the reader may encounter in the literature are summarized. Secondly, spatial differentials of length, area, and volume are examined. Finally, the concept and definition of partial and total derivatives are given for scalar as well as vector functions. In this latter regard, the idea that derivatives of unit vectors must in general take into account changes in direction and therefore may not be zero is developed for later use.

## 1.1 Notation

A consistent notation, which we will refer to as *explicit standard notation*, that can be used for handwritten or electronic communication between researchers, innovators, designers, and academics (including, of course, students and instructors) is suggested. Therefore, this notation eschews the use of boldface that is common in the literature for denoting quantities that have direction, such as vectors. Scalars, vectors, dyadics, and other tensors, as well as phasors, are cited in explicit standard notation in Sections 1.1.1 through 1.1.7(a) below. Explicit standard notation uses the multiple overbar to denote tensors of varying *rank*. Rank is a property of a quantity that signifies *directional compoundedness*—a term that will be used throughout this guide. This multiple overbar notation is in frequent use in current texts in fields and photonics.[1]

Another common notation called *tensor notation*,[2] which uses multi-subscripts to denote multiple directivity of tensors, is listed in Section 1.1.7(b). Tensor notation is perhaps the most thorough because the ordering of its subscripts denotes the internal structure of the tensor that it depicts. For that reason, tensor notation is used in this guide whenever appropriate.

Various other notational representations that the user may encounter are listed in Sections 1.1.7(c) through 1.1.7(e). Finally, the description of another notation called *order notation*, which is also in common use[3–7] is postponed until Chapter 3 because it is quite another matter.

### 1.1.1 Scalars

Quantities that have a magnitude but no directionality are *scalars* and can be written as, for example,

$$a, A, 1, \text{ or } 27,$$

without any over- or undermarkings. Quantities such as pressure, temperature, voltage (electric potential), entropy, and work or energy are scalars because they have no directionality, only magnitude. Other quantities, such as components of vectors, current, and flux, may vary with coordinate rotations—a topic discussed in Chapter 2—but are also scalars by definition.

### 1.1.2 Vectors

A quantity that has a magnitude and an inherent *single** direction is referred to as a *vector*. In explicit standard notation, a 3-space vector, that is, a vector quantity in three-dimensional (3D) space, having a magnitude $a$ is written with an overbar and is commonly expanded into three orthogonal coordinates, such as

$$\overline{a} = \hat{u}_x a_x + \hat{u}_y a_y + \hat{u}_z a_z \qquad (1.1\text{-}1)$$

in Cartesian coordinates, where $a_x$ is the component of $\overline{a}$ in the $x$ direction, $a_y$ is the component of $\overline{a}$ in the $y$ direction, $a_z$ is the component of $\overline{a}$ in the $z$ direction, and where $\hat{u}_x$, $\hat{u}_y$, and $\hat{u}_z$ are unit vectors aligned along orthogonal Cartesian coordinate axes (in this case). Unit vector notation with the hat overmarkings is discussed in Section 1.1.3. In many texts, lowercase vowels $a$, $e$, $i$, or $u$ are reserved for unit vectors. In Eq. (1.1-1), "$\overline{a}$" is used in contrast as a full vector with all three components, $a_x$, $a_y$, and $a_z$. In other books, $\boldsymbol{a}_x$, $\boldsymbol{a}_y$, and $\boldsymbol{a}_z$ in bold lowercase represent unit vectors.[8–10] Thus, by using the overbar exclusively for vectors of general magnitude and the "u-hat" for unit vectors, the door is open for multiple overmarkings to denote dyadics and other tensors.

Another vector having a magnitude $A$ will contain the overbar as before and, if expanded in generalized coordinates, is written as

---

* The word *single* is emphasized here because vectors are not the only quantities that have direction. As we will soon learn, quantities can have multiple directionality and, thus, are not vectors. But, since vectors inherently have only one direction, we insert the word single to make that distinction.

$$\overline{A} = \hat{u}_1 A_1 + \hat{u}_2 A_2 + \hat{u}_3 A_3 \tag{1.1-2}$$

where subscripts 1, 2, and 3 represent three distinct orthogonal directions in 3-space. Quantities such as force, velocity, acceleration, flow field, current density, electric and magnetic field intensities, flux densities, and polarization fields each have a magnitude and an inherent single direction. These are therefore expressed as vectors.

Vector components are inherently scalars but can be made to show direction by including their associated unit vector. For example, the scalar $A_1$ above may be depicted as a vector by attaching a unit vector to it: $\overline{A_1} = \hat{u}_1 A_1$. Current $I$ and flux $\psi$, which are scalars by definition, are often vectorized by incorporating a unit vector in their assumed direction. Although care must be taken when doing this, current and flux as vectors would be written as $\hat{u}_i I$ and $\hat{u}_i \psi_i$ or just $\overline{I}$ and $\overline{\psi}$. However, it is generally better to work with the densities, such as current density or flux density, which are inherently vector fields and are given units of the flux per meter squared (in SI units).* For example, current density $\overline{J}$ is the flux density of current $I$ (amps), and is given in units of amps per square meter.

### 1.1.3 Unit vectors

A *unit vector* is a quantity that has a magnitude of *one* and has an inherent single direction associated with it. In our explicit standard notation, a unit vector is written as a "$u$" with a hat or single chevron overmarking, such as

$$\hat{u}_i$$

where the subscript is used to indicate a direction—in this case, the $i$th direction. Unit vectors are not confined to coordinate directions, although it is common to

---

* The standard of units called the "SI" system is now managed by the U.S. National Institute of Standards and Technology (NIST). See special publication #330, 1971. In 1907, it was first proposed by Giorgi, who named it MKS (for meter-kilogram-second) and later, rationalized MKS. SI is the acronym for *Système internationale d'unités*. Whereas the SI system is the common standard in engineering literature, it is becoming prevalent in the scientific literature as well. Nevertheless, the reader should become familiar with the "Gaussian" system as well. The Gaussian system is a combination between the earlier *emu* (electromagnetic, meaning magnetostatic, units) and *esu* (electrostatic units) systems. It has a mathematical purity that renders the electric and magnetic fields in Maxwell's equations similar in form and units, for example.

do so. Other symbols (usually lowercase) are often used for unit vectors. However, the author recommends the above redundant use of both the single chevron and the lowercase $u$ in student-instructor exchanges to make it clear that a unit vector is intended.

### 1.1.4 $\overline{r}$ - space notation: the vector-like $\overline{r}$ used in the argument of a field function

It is common in mathematical literature of science and engineering to see field functions that contain a vector-like looking symbol $\overline{r}$ in their functional arguments. We will refer to this usage as *$\overline{r}$-space notation.* For example, the electric potential $V$ at a point in space called the *field point* due to a system of charges denoted by $\rho$ located at *source points* would typically be written as

$$V(\overline{r}) = \frac{1}{4\pi\varepsilon}\int_{Sources} \frac{\rho(\overline{r}\,')}{\left|\overline{r} - \overline{r}\,'\right|} dv' \tag{1.1-3}$$

where the volume is taken over all of the positions where sources are present.

Here the symbol $\overline{r}$ in the argument of $V$ represents a shorthand notation for the spatial position of the field point, where the unknown electric potential field is being determined. Likewise, the $\overline{r}\,'$ in the argument of $\rho$ represents the spatial position of the known charge sources. The integration is being taken over the volume elements $dv'$ where the charge sources are known to be located. The prime denotes source positions while the lack of a prime denotes field positions. Thus, whenever the symbols $\overline{r}$ or $\overline{r}\,'$ are used in the argument of a function, it is meant simply as a shorthand for coordinates separated by commas or is used where no coordinates are implied at all. For example, in Cartesian coordinates, $\overline{r}$ and $\overline{r}\,'$ are $x, y, z$ and $x', y', z'$. Likewise, in cylindrical or spherical coordinates, $\overline{r}$ and $\overline{r}\,'$ are $r, \phi, z$ and $r', \phi', z'$ or $r, \theta, \phi$ and $r', \theta', \phi'$, respectively. For generalized coordinates, $\overline{r}$ and $\overline{r}\,'$ are $q_1, q_2, q_3$ and $q_1', q_2', q_3'$. This is what we mean by *$\overline{r}$-space notation.* The symbols $\overline{r}$ and $\overline{r}\,'$ in the argument of a function represent a point in space, with or without regard to a coordinate system.

However, having said that, $\overline{r}$ and $\overline{r}\,'$ do, in fact, have a physical interpretation. See Fig. 1.1-1. In the case of $\overline{r}$, for example, the symbol can represent the vector from an arbitrarily selected point in space designated as the origin $O$ to the point $P$ in space where the field $V$ is to be determined. In the case

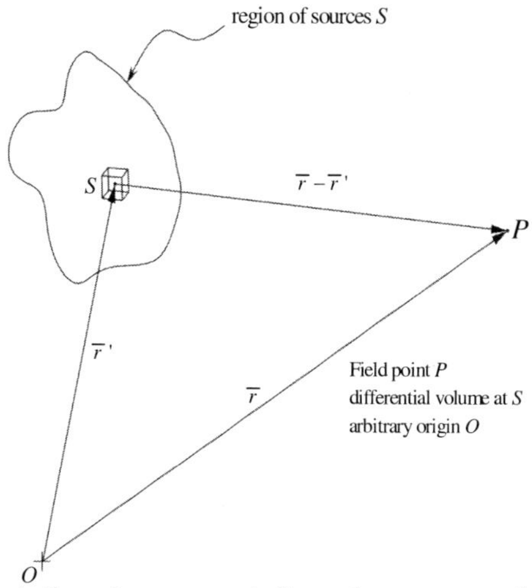


**Figure 1.1-1** Notational representation of source and field points in $\overline{r}$ *– space notation.*

of $\overline{r}'$, this symbol represents the vector from the same origin to the source points *S*, over which the integration takes place. In both cases, however, this terminology is used independently of any specific coordinate system. We will refer to fields written this way as *fields in* $\overline{r}$ *space*. The vector drawn from the source element to the field point is $\overline{r}-\overline{r}'$. This can be written as capital *R* vector, so that $\overline{R}=\overline{r}-\overline{r}'$. See Fig. 1.1-1 for the geometry associated with $\overline{r}$, $\overline{r}'$, and $\overline{r}-\overline{r}'$.

### 1.1.5 Phasors

In engineering and physics, quantities frequently vary in time as well as space. Whenever quantities vary sinusoidally in time at a monochromatic frequency, they are referred to as *time harmonic*. Further, it is customary to depict the time-harmonic time variation as

$$e^{-i\omega t} \text{ or } e^{j\omega t}$$

where $\omega$ is the monochromatic radian frequency. In physics, the first exponential is commonly used, whereas in engineering, the second one is the more common. There is no substantive advantage of one over the other and one can easily convert analyses done in one to that of the other by replacing $j$ with $-i$ or vice versa.

A time-harmonic scalar quantity $f(x,y,z,t)$ would then be written as

$$f(x, y, z, t) = \underline{f}(x, y, z)\, e^{j\omega t} \tag{1.1-4}$$

The remaining analysis is then performed with the exponential factor suppressed. The function

$$\underline{f}(x, y, z)$$

is then referred to as a *phasor* and can be denoted as such by the underscore either in electronic or handwritten communication. Any quantity that is dealt with in this manner with the time exponential suppressed, whether a scalar, vector, or other, is referred to as a *phasor*. The above function $\underline{f}$ is then a *scalar phasor*.

A time-harmonic time-varying vector would be written as

$$\overline{A}(x, y, z, t) = \underline{\overline{A}}(x, y, z)\, e^{-i\omega t} \tag{1.1-5}$$

where $\underline{\overline{A}}(x, y, z)$ is a *vector phasor* denoted by an overbar as well as an underbar. Phasor scalars and phasor vectors are discussed in greater detail in Section 2.5. Dyadics, which are discussed in Section 1.1.6 below and in Chapter 3, can also be phasors, as can tensors in general. See Section 2.5 for *tensor phasor* notation.

### 1.1.6 Dyadics

One's first step in understanding dyadics is to overcome the misconception that all quantities either have *a* direction or not. The emphasis here is on the word *a*, meaning a *single* direction or not. A student learning about vector fields finds that they are not overly difficult to visualize once scalar fields are understood. Vector fields are commonly described as quantities that have a direction (as well as magnitude) at every point in space (as well as at every moment in time). Such quantities are spatial and temporal. Implied in that first exposure to vectors was the word "single," meaning single direction, even though this distinction was probably not mentioned at the time.

The extension of the concept of vector fields to another type of quantity called *dyadics* is to understand that quantities can have *two* directions at every point in space and at every moment in time. Some examples below will aid in the understanding of this concept. We will refer to such quantities as having *dual directional compoundedness* and, as we just stated, name such quantities as *dyadics*. This dual directivity should not be construed as a simple combination of two three-component vectors. Rather, each of the three components of one vector acts linearly on each of the three components of the other yielding nine components that fully describe the dyadic. That is,

> *each of the nine components of the dyadic has a magnitude and a dually directed unitary dyadic called a unit dyad just as does each component of a vector have a magnitude and a singly directed unitary vector called a unit vector.*

While the following discussion is focused on notation and most of the examples in this section are for quantities with dual directional compoundedness, the reader should keep in mind that tensors in general can have multiple directional compoundedness beyond two. The general term for such quantities, where the multiplicity of the directional compoundedness is unspecified, is *tensor*. In Section 1.1.7, we list several common notational representations for tensors to prepare the reader for more detailed discussions in subsequent chapters about tensors and special cases of tensors, such as dyadics.

Although junior-level (third-year) engineering or physics students have not necessarily been required to take a formal course in tensor analysis, they frequently have been exposed to quantities that have multiple directionality, often without being told so. However, in recent years more attention is being devoted to dyadics and other tensors because of the recognition of their importance, even at this level.

### *Example 1: piezoelectric transducers and other crystalline materials*

In the study of anisotropic dielectrics, the dielectric properties cannot be fully described without the use of dyadics. That is, a dyadic is needed to express the constitutive relation between the electric flux in a crystalline dielectric and the applied electric field. For such materials, the flux density vector is not necessarily parallel with the applied electric field intensity vector because the crystalline material has different dielectric properties in different directions. Thus, the dielectric must be described by a dyadic. This feature has a practical and necessary application not only to crystalline dielectrics, but also to piezoelectric materials commonly used for sonar transmitters and receivers. The piezoelectric phenomenon found in certain special anisotropic materials is used in the design of transducers to convert between acoustic waves and electric fields.

### *Example 2: Magnetostrictive transducers*

Another example of the need for dyadics is in the use of magnetostrictive materials, which serve as converters between acoustic waves and magnetic fields. In these materials, the magnetic properties must be described by a dyadic in order to properly account for the magnetic flux density vector field that results from an applied magnetic field intensity vector field.

### *Example 3: Stress and strain mechanics of materials*

Static and dynamic analyses of the mechanics of materials are typical in many engineering curricula. The stress and strain quantities covered in these courses are, in fact, dyadics. Each of these dyadic entities serves as a transformation between two vector fields that are not necessarily parallel in the material. Because these courses are usually taken at the sophomore (second-year) level, the tensorial nature of these quantities is usually not mentioned.

### *Example 4: Conversion between coordinate systems*

In Appendix B, various $3 \times 3$ matrix transformations between orthogonal coordinate systems (in 3D space) are presented along with certain other coordinate system properties. These nine-component transformations between coordinate systems can be, and often are, viewed as dyadics.

### *Example 5: The Jacobian differential operator*

Yet another example of the utility of dyadics is in numerical analysis. When studying Newton's method of locating roots of systems of nonlinear equations, students come across the Jacobian matrix. The Jacobian is the determinant of the resultant components of an $n$-dimensional vector differential operator operating on an $n$-dimensional vector. It therefore has a dual directionality—one implicit from the vector operator (which we will later discover is the gradient operator) and one given by the vector upon which it operates. It is described by $n^2$ components and is therefore not a scalar or vector, yet it has magnitude and direction. The key here is that it has two directions.

Quantities that have dual directionality are referred to as *dyadics* as stated earlier. Philosophically, this duality of direction can be considered from a causality viewpoint, that is, cause-and-effect. The cause is a forcing function, which can have direction, and the effect is the resultant, which can also have direction, but not necessarily parallel to the causal direction. Thus, at every point in space and at every moment in time, there are causal and resultant directions. *The transformation matrix between these two quantities* is *the dyadic, if the causal and resultant quantities are vector fields*. The mathematical expression relating resultant field to the causal field through the dyadic matrix transformation is referred to as a *constitutive relationship*.

(Having said that, we hasten to point out the *if* in the italicized sentence above, for such constitutive relationships are not always fully describable by dyadic tensors. A case will be shown in Chapter 3 where the causal and resultant fields are themselves the stress and strain dyadics mentioned in Example 3. In this case, the constitutive relation involves a tensor having a four-level directional compoundedness. Therefore, the tensor that describes this is not a dyadic but is a rank-four tensor. It may surprise the student contemplating such a concept for the first time that we are talking about the elastic modulus. This oh-by-the-way parenthetical comment is made at this point in our discussion as a caution to the reader *not to think of tensors as just being dyadics*. Dyadics are special cases of tensors, but tensors are not necessarily dyadics. This point will be restated at various strategic places in this mathematical guide.)

> *Rank: The quantitative property of a tensor that specifies its directional compoundedness is "rank." Thus, dyadics are tensors of "rank two," because of their dual directivity. Similarly, vectors and scalars are also tensors but of rank one and zero, respectively, because vectors have single directivity and scalars have no directivity.*

We will discuss dyadics and other tensors of higher directional compoundedness (higher rank) in more detail in Chapter 3. For now, since our purpose here is notation, we will denote the dual directionality of dyadics with a double overbar, such as

$$\bar{\bar{A}} \tag{1.1-6a}$$

and its associated unit dyad as a $u$ with a double hat or double chevron, such as

$$\hat{\hat{u}} \tag{1.1-6b}$$

where each has dual directionality. In 3D space, therefore, these quantities have nine components, as previously stated.

In this guide, we have been referring to the notation of using overbars, underbars, and chevrons, such as in Eqs. (1.1-1) through (1.1-6b), as *explicit standard notation.* Many other notations are summarized below so as to acquaint the reader with various other forms encountered in the literature.

### 1.1.7 Tensors

Having just taken the "uninitiated" (to tensors) through the concept of dyadics, which are rank-two tensors, we will leave further extension of this concept to quantities having multiple directional compoundedness greater than two to Chapter 3, Section 4.7, and Chapter 5. Such quantities in general are called *tensors*, as stated earlier. Various authors use a variety of notational techniques for denoting tensors. One of the most common is the multiple-subscript method called *tensor notation.*

> *Tensor notation has a great deal of utility in that it explicitly allows for the proper ordering of tensor components in performing various operations, such as single or multiple dot-, cross-, or direct-product operations. Therefore, tensor notation is the preferred formulation whenever these operations are important to the development at hand.*

Other notational techniques for representing tensors (besides explicit standard notation and tensor notation) that may be encountered in the literature include the use of pre-subscripts, pre-superscripts, various arrow overbar symbols, as well as post-subscript and post-superscript methods. Finally, there is the so-called *order method.* Each method has its own utility with its respective advantages and disadvantages.

Explicit standard notation is illustrated for rank-three tensors in Section 1.1.7(a) below. Tensor notation is denoted in Section 1.1.7(b), pre-subscript/superscript methods in Section 1.1.7(c), and the double-pointed arrow in Section 1.1.7(d). Post-subscript and post-superscript methods for depicting tensors are mentioned in Section 1.1.7(e).

In the following synopsis of notation used for tensors, the examples are given first for dyadics and then for *triadics*, the latter being added for generality, so that the newcomer to tensors (as well as those with a restricted understanding of tensors) will garner the impression that there are tensors beyond dyadics. This is done without regard for what exactly dyadics and triadics actually are. The what-are-dyadics-and-triadics discussion will come later. For now, we are just dealing with the names of things and their notations.

Finally, the *order method*, which is common in recent usage in nonlinear optics,[3–7] is discussed in Section 3.1. For the reasons discussed at the end of the preceding paragraph, the notational approach for the order method is not included here, as it differs somewhat from the other methods.

### 1.1.7(a) Explicit standard notation for tensors

In Section 1.1.6 we cited $\overline{\overline{A}}$ and $\hat{\hat{u}}$ as the notational representations of the dyadic and unit dyad as shown in Eqs. (1.1-6a) and (1.1-6b). This is what we call *explicit standard notation* for quantities that have dual directional compoundedness.

Quantities that have triple directional compoundedness are called *triadics*. In explicit standard notation, triadics are given the notational representation of three overbars, as

$$\overline{\overline{\overline{A}}} \tag{1.1-7a}$$

and their associated unit *triads* as a *u* with a triple chevron:

$$\hat{\hat{\hat{u}}} \tag{1.1-7b}$$

This process, of course, would continue with further increase in directional compoundedness. (See Appendix C.)

### 1.1.7(b) Multiple-subscript notation for tensors

It is often more convenient to use *tensor notation* when expressing quantities having multiple directivity. Tensor notation is a method of denoting tensors with a series of integer subscripts called *indices*. The number of indices corresponds to the directional compoundedness level of the tensor quantity, which will be covered in Section 3.1. Therefore, our dyadic $\overline{\overline{A}}$ and unit dyad $\hat{\hat{u}}$ of expressions (1.1-6a) and (1.1-6b), which are rank-two tensors, would appear simply with double indices when written in (multisubscript) *tensor notation*, as

$$A_{ij} \tag{1.1-8a}$$

and

$$u_{ij} \tag{1.1-8b}$$

respectively. Notice that it is not necessary to use the double hat (chevron), since the double indices denote the rank-two unit dyadic.

In tensor notation vectors would be written with a single index, such as $A_i$. However, since the symbol $A_i$, written to represent a vector in tensor notation, cannot be distinguished from $A_i$ as the *i*th scalar component of that vector, it is necessary to clarify whenever tensor notation is being used in order to make the distinction.

Triadics, or rank-three tensors, are written with three indices such as

$$T_{ijk} \tag{1.1-9a}$$

and the unit *triad* as

$$u_{ijk} \tag{1.1-9b}$$

Again, the unit triad may or may not contain chevrons.

### 1.1.7(c) Pre-subscript, pre-superscript notation for tensors

The dyadic and *unit dyad* of expressions (1.1-6a) and (1.1-6b) are expressed in pre-subscript notation as

$${}_2A \text{ or } \left[{}_2A\right] \tag{1.1-10a}$$

and

$${}_2u \text{ or } \left[{}_2u\right] \tag{1.1-10b}$$

respectively, where the pre-subscript denotes the directional compoundedness of the quantity. The unit dyads may or may not be wearing a hat or a double chevron; however, such redundancy is not necessary when the pre-subscript is used as shown in expressions (1.1-10a) and (1.1-10b).

In pre-subscript notation, triadics would appear as

$${}_3T \text{ or } \left[{}_3T\right] \tag{1.1-11a}$$

and the *unit triad* as

$$ {}_{3}u \text{ or } \left[ {}_{3}u \right] \qquad (1.1\text{-}11b) $$

Alternatively, the dyadic and the unit dyad of expressions (1.1-6a) and (1.1-6b) are expressed in pre-superscript notation as

$$ {}^{2}A \text{ or } \left[ {}^{2}A \right] \qquad (1.1\text{-}12b) $$

$$ {}^{2}u \text{ or } \left[ {}^{2}u \right] \qquad (1.1\text{-}12b) $$

respectively, where the pre-superscript denotes the directional compoundedness and the unit dyads may or may not have a hat or chevrons.

In pre-superscript notation triadics would appear as

$$ {}^{3}T \text{ or } \left[ {}^{3}T \right] \qquad (1.1\text{-}13a) $$

and the unit triad as

$$ {}^{3}u \text{ or } \left[ {}^{3}u \right] \qquad (1.1\text{-}13b) $$

again, where the unit triads may redundantly be given the chevron.

### 1.1.7(d) Arrow notation for tensors

The use of arrows is another method for denoting quantities that have directionality. In this notation, the vector appears with an overarrow having a single arrowhead as

$$ \vec{A} \qquad (1.1\text{-}14) $$

and the dyadic appears with an arrow having arrowheads at both ends as

$$ \overleftrightarrow{A} \qquad (1.1\text{-}15) $$

The overarrow can be split with a number inserted indicating the directional compoundedness. Thus for a triadic, the overarrow would have a 3 inserted in the overarrow as

$$\overset{\leftarrow 3 \rightarrow}{T} \tag{1.1-16}$$

Unit vectors, unit dyads and unit triads in this notation either use the chevrons or the arrows as in expressions (1.1-14), (1.1-15) and (1.1-16) with the lowercase $u$ or other vowel.

### 1.1.7(e) Post-subscript, post-superscript notation for tensors

The reader should also be aware of post-subscripts and post-superscripts being used to denote tensor directional compoundedness. Post-subscripts are commonly used to distinguish one quantity from another. For example, using the notation of expression (1.1-10a), the dyadic $[{}_2A_1]$ could be distinguished from another dyadic $[{}_2A_2]$ by the use of the post-subscript in the same way that vector $\bar{A}_1$ might be distinguished from another vector $\bar{A}_2$. In the case of post-superscripts, there is possible confusion with degrees of power, such as the squaring or cubing of a quantity.

Nevertheless, authors use both of these methods to denote tensor directional compoundedness. In each case, however, authors are usually careful to specify at the outset what is meant by such notation in order to distinguish it from these other uses. Although some have parenthesized post-subscripting, the use of parentheses in post-superscripting is currently reserved (by recent common usage)[3–7] to order notation. In any case, authors are careful to specify their notation, and the reader new to tensors should watch carefully for this in order to avoid ambiguity in interpretation of the author's meaning.*

At the risk of having left out someone's favorite notation, we have endeavored to cite some of the common notations used in the tensor literature. The reader should realize, however, that there are no standards for these things, and that authors have the freedom to set up any notation that they please.

## 1.2 Spatial Differentials

Differential lengths, areas, and volume are discussed in the subsections that follow.

---

* The use of the post-superscript to denote tensor rank is, however, sufficiently common that it is felt worthy of mention, if for no other reason then to caution those new to tensors to such use.

### 1.2.1 Differential length vectors

Differential lengths are the building blocks to differential areas (Section 1.2.2) and differential volumes (Section 1.2.3). For this reason the tables in Appendix B focus on differential lengths. These tables provide explicit differential lengths for several orthogonal coordinate systems. Perspective views of orthogonal coordinate surfaces graphically depict the product of these building blocks.

A differential length $\overline{d\ell}$ in any orthogonal coordinate system can be written as the vector

$$\overline{d\ell} = \hat{u}_1 d\ell_1 + \hat{u}_2 d\ell_2 + \hat{u}_3 d\ell_3 \qquad (1.2\text{-}1)$$

or alternatively, as a summation over the dimensionality of the space, such as

$$\overline{d\ell} = \sum_{i=1}^{3} \hat{u}_i \, d\ell_i \qquad (1.2\text{-}2)$$

for 3D space, where $\hat{u}_1, \hat{u}_2, \hat{u}_3$ are three orthogonal unit vectors in that space. The unit for differential length in the SI system is meters.

As stated earlier, vectors are a special case of tensors. It is often more convenient, especially when working with generalized tensors in conjunction with vectors, to omit (or suppress) the details of Eqs. (1.2-1) and (1.2-2) for simplicity in notation. Using such tensor notation, our differential length vector may be written simply as

$$d\ell_i \qquad (1.2\text{-}3)$$

where the summation, unit vectors, and overbar are implied. It is necessary to clarify when tensor notation is being used in order to distinguish, for example, the difference in the meaning of $d\ell_i$ in Eq. (1.2-2) and in the expression (1.2-3). In (1.2-2), $d\ell_i$ represents the scalar magnitudes of the components of $\overline{d\ell}$, whereas in (1.2-3), $d\ell_i$ ***is*** the vector $\overline{d\ell}$.

### 1.2.2 Differential area

A differential area may be defined as the area subtended by two orthogonal differential lengths as illustrated in Fig. 1.2-1. It is given by

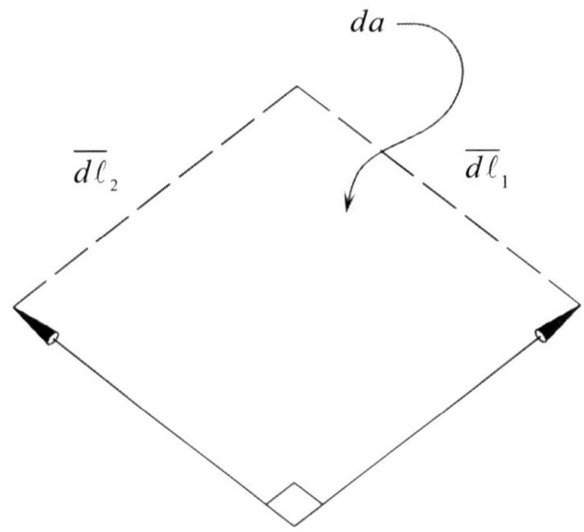


**Figure 1.2-1** Differential area as a scalar.

$$da = |d\ell_1||d\ell_2| \tag{1.2-4}$$

which is a scalar. Its (SI) units are meters squared.

However, differential area can be given a direction. Two vectors not in the same direction define a surface. Thus, a unit vector $\hat{u}_n$ normal to that surface may be constructed as shown in Fig. 1.2-2. Notice that there are two possible unit normals, one shown by the solid line and one by the dashed line in Fig. 1.2-2. Thus, a convention must be established to remove this two-to-one ambiguity. The one that is chosen depends on the nature of the problem. The usual convention is to use the right-hand rule by pointing the fingers of the right hand in the direction of $\hat{u}_1$ and then rolling them into the direction $\hat{u}_2$. The thumb will then point in the "normal" direction. On a closed surface, it is customary to construct $\hat{u}_1$ and $\hat{u}_2$ such that the normal will be directed outward. The important point to remember here is that differential area is inherently a vector that can be represented by

$$\overline{da} = \hat{u}_n |d\ell_1||d\ell_2| \tag{1.2-5}$$

Conveniently, the unit normal given in Eq. (1.2-5) can be expressed in terms of the cross product between $\overline{d\ell}_1$ and $\overline{d\ell}_2$, as covered in Eq. (2.4-30). Therefore, vector differential area is also given another useful form in that section.

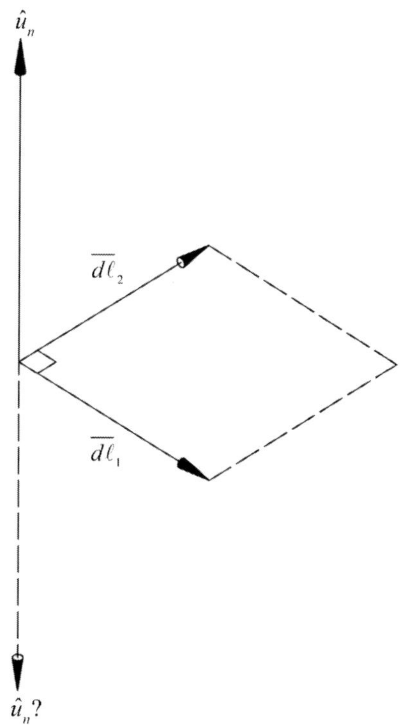


**Figure 1.2-2** Differential area as a vector.

### 1.2.3 Differential volume

Three orthogonal differential lengths $d\ell_1, d\ell_2, d\ell_3$ define a differential volume $dv$. Given three generalized orthogonal coordinates $q_1, q_2, q_3$ as shown in Fig. 1.2-3, differential volume is the generalized volume parallelepiped shown. It is

$$dv = |d\ell_1|\,|d\ell_2|\,|d\ell_3| \tag{1.2-6}$$

in meters cubed (SI units). Because there is no directionality associated with a mathematical differential volume, $dv$ is a scalar.

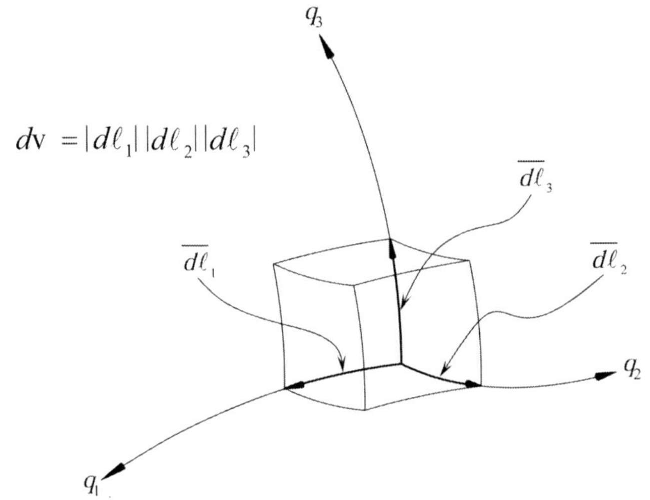


**Figure 1.2-3** Differential volume.

## 1.3 Partial and Total Derivatives

Since the vector differential operators we deal with in the mathematics of fields and photonics (especially in Chapter 4) consist of various combinations of partial derivatives of scalars, vectors, dyadics and, in general, tensors, a short review of partial derivatives is in order.

> *Definition of a partial derivative: A partial derivative is the result of taking a derivative of a function of multiple independent variables with respect to one of the variables while holding all of the other independent variables constant.*

Therefore, the partial derivative represents the rate of change of a function of multiple variables only along one of the variables, that is, while the other variables are unchanged. We first discuss the case of a partial derivative of a scalar function and then do the same for vector fields, ending with a mention of extending this to higher-rank tensors.

In the event that the total rate of change is required allowing all of the variables to change, the total derivative is needed. We therefore examine the chain rule for determining the total derivative of a multivariate function. We shall see that the chain rule requires the partial derivatives with respect to each of the variables.

### 1.3.1 Partial derivative of a scalar function

The partial derivative of the scalar function $f(q_1,q_2,q_3)$ with respect to $q_1$ is defined as

$$\frac{\partial f}{\partial q_1} = \lim_{\Delta q_1 \to 0} \frac{f(q_1+\Delta q_1,q_2,q_3)-f(q_1,q_2,q_3)}{\Delta q_1} \tag{1.3-1}$$

which is often written in shorthand simply as

$$\frac{\partial f}{\partial q_1} = f_{q_1} \tag{1.3-2}$$

Likewise, the partial derivative of $f(q_1,q_2,q_3)$ with respect to $q_2$ is

$$\frac{\partial f}{\partial q_2} = \lim_{\Delta q_2 \to 0} \frac{f(q_1,q_2+\Delta q_2,q_3)-f(q_1,q_2,q_3)}{\Delta q_2} \tag{1.3-3}$$

which can be written more conveniently as

$$\frac{\partial f}{\partial q_2} = f_{q_2} \tag{1.3-4}$$

Finally, the partial derivative of $f(q_1,q_2,q_3)$ with respect to $q_3$ is

$$\frac{\partial f}{\partial q_3} = \lim_{\Delta q_3 \to 0} \frac{f(q_1,q_2,q_3+\Delta q_3)-f(q_1,q_2,q_3)}{\Delta q_3} \tag{1.3-5}$$

which can be expressed as

$$\frac{\partial f}{\partial q_3} = f_{q_3} \tag{1.3-6}$$

Second-order partial derivatives are defined by

$$\frac{\partial^2 f}{\partial q_1^2} = \frac{\partial f}{\partial q_1}\left(\frac{\partial f}{\partial q_1}\right) = f_{q_1 q_1} \tag{1.3-7}$$

$$\frac{\partial^2 f}{\partial q_1 \partial q_2} = \frac{\partial f}{\partial q_1}\left(\frac{\partial f}{\partial q_2}\right) = f_{q_1 q_2} \tag{1.3-8}$$

$$\frac{\partial^2 f}{\partial q_2^2} = \frac{\partial f}{\partial q_2}\left(\frac{\partial f}{\partial q_2}\right) = f_{q_2 q_2} \tag{1.3-9}$$

$$\frac{\partial^2 f}{\partial q_2 \partial q_1} = \frac{\partial f}{\partial q_2}\left(\frac{\partial f}{\partial q_1}\right) = f_{q_2 q_1} \tag{1.3-10}$$

***The mixed derivative theorem*** *states that if* $f(q_1, q_2, q_3)$ *and its partial derivatives* $f_{q_1}$, $f_{q_1}$, *and* $f_{q_1 q_2}$ *exist and are continuous, then* $f_{q_2 q_1}$ *also exists and*

$$f_{q_1 q_2} = f_{q_2 q_1} \tag{1.3-11}$$

## 1.3.2 Total derivative of a scalar function: chain rules

First, we describe the total differential $df$ as

$$df = \frac{\partial f}{\partial q_1} dq_1 + \frac{\partial f}{\partial q_2} dq_2 + \frac{\partial f}{\partial q_3} dq_3 \tag{1.3-12}$$

This represents the change produced in $f$ by changes in $q_1$, $q_2$, and $q_3$.

### 1.3.2(a) Chain rule for functions of three independent variables

If $f = f(q_1, q_2, q_3)$ and $q_1$, $q_2$, and $q_3$ are functions of an independent variable $t$ (and all are differentiable), then the total derivative of $f$ with respect to $t$ is

$$\frac{df}{dt} = \frac{\partial f}{\partial q_1}\frac{dq_1}{dt} + \frac{\partial f}{\partial q_2}\frac{dq_2}{dt} + \frac{\partial f}{\partial q_3}\frac{dq_3}{dt} \tag{1.3-13}$$

### 1.3.2(b) Chain rule for surface functions

If $f = f(q_1, q_2, q_3)$ is a differentiable scalar function that is confined to a surface (for example, the surface of a sphere) described by two independent variables $\theta$ and $\phi$ at a constant $r = r_o$, then $q_1$, $q_2$, and $q_3$ are functions of $\theta$ and $\phi$ (and all are differentiable), and the partial derivatives of $f$ with respect to $\theta$ and $\phi$ are

$$\frac{\partial f}{\partial \theta} = \frac{\partial f}{\partial q_1}\frac{\partial q_1}{\partial \theta} + \frac{\partial f}{\partial q_2}\frac{\partial q_2}{\partial \theta} + \frac{\partial f}{\partial q_3}\frac{\partial q_3}{\partial \theta} \tag{1.3-14}$$

and

$$\frac{\partial f}{\partial \phi} = \frac{\partial f}{\partial q_1}\frac{\partial q_1}{\partial \phi} + \frac{\partial f}{\partial q_2}\frac{\partial q_2}{\partial \phi} + \frac{\partial f}{\partial q_3}\frac{\partial q_3}{\partial \phi} \tag{1.3-15}$$

## 1.3.3 A dimensionally consistent formulation of partial derivatives

Whereas Eqs. (1.3-1) through (1.3-15) provide the mathematical definitions of partial derivatives of the scalar function $f$ with respect to the generalized orthogonal coordinates, $q_1$, $q_2$, and $q_3$, it should be recognized that these coordinates are not necessarily dimensionally consistent. Some may be in units of length (meters) while others are in units of angle (radians), which are unitless. Therefore, in order for all of the partial derivatives to be dimensionally consistent, *scale factors* (otherwise known as *metric coefficients*) are used. These factors are discussed in greater detail in Section 2.6, but for now we will just cite an example of how they are used for cylindrical coordinates.

The scale factors $h_1$, $h_2$, and $h_3$ are used to relate the differential length components $d\ell_i$ in Eq. (1.2-1) to the differential coordinates $dq_i$ by $d\ell_i = h_i dq_i$. In cylindrical coordinates, $q_1$, $q_2$, and $q_3$ are $r$, $\phi$, and $z$. Thus, $h_1$, $h_2$, and $h_3$ are 1, $r$, and 1, respectively, and $d\ell_1 = dr$, $d\ell_2 = r d\phi$, and $d\ell_3 = dz$. Notice that each of the differentials has units of length once the scale factor $r$ is included in the second differential length.

In this way a dimensionally consistent formulation for partial derivatives in generalized orthogonal curvilinear coordinates (GOCCs) may be constructed. Therefore,

$$\frac{\partial f}{h_i \partial q_i} = \lim_{\Delta q_i \to 0} \frac{f(q_i + \Delta q_i, q_{i+1}, q_{i+2}) - f(q_i, q_{i+1}, q_{i+2})}{h_i \Delta q_i} \tag{1.3-16}$$

A surface in 3D space given in generalized orthogonal curvilinear coordinates $q_1, q_2, q_3$ is described as $h_3 q_3 = f(h_1 dq_1, h_2 dq_2)$. Then the change in $f$ with respect to $d\ell_1 = h_1 dq_1$, for example, is

$$\frac{\partial f}{d\ell_1} = \frac{\partial f}{h_1 \partial q_1} = \frac{\partial (h_3 q_3)}{h_1 \partial q_1} = \frac{h_3}{h_1}\frac{\partial q_3}{\partial q_1} + \frac{q_3}{h_1}\frac{\partial h_3}{\partial q_1} \tag{1.3-17}$$

where the metric coefficients are in general functions of all three coordinates $h_i = h_i(q_1, q_2, q_3)$ and therefore must be included in the derivative when applying the product rule.

### 1.3.4 Partial derivative of a vector function

By the time students in the physical sciences or engineering enter upper-division courses (junior and senior years of a bachelor's program), they will have been exposed to the partial derivative. However, this introduction was invariably done in Cartesian coordinates with Cartesian-coordinate examples. This was fine when the partial derivative being explained was taken on a scalar function. However, problems can arise if the partial derivative is taken of a vector function (or of any tensor of rank greater than zero) and the physical problem leads naturally into curvilinear coordinates, such as cylindrical coordinates. We will therefore discuss partial derivatives of vectors in generalized orthogonal curvilinear coordinates.

Let us examine the partial derivative of a vector field $\bar{A}(q_1, q_2, q_3) = \hat{u}_1 A_1(q_1, q_2, q_3) + \hat{u}_2 A_2(q_1, q_2, q_3) + \hat{u}_3 A_3(q_1, q_2, q_3)$. The partial derivative of $\bar{A}$ with respect to one of the coordinates $q_i$ is

$$\begin{aligned} \frac{\partial \bar{A}}{\partial q_i} &= \frac{\partial (\hat{u}_1 A_1)}{\partial q_i} + \frac{\partial (\hat{u}_2 A_2)}{\partial q_i} + \frac{\partial (\hat{u}_3 A_3)}{\partial q_i} \\ &= \hat{u}_1 \frac{\partial A_1}{\partial q_i} + \hat{u}_2 \frac{\partial A_2}{\partial q_i} + \hat{u}_3 \frac{\partial A_3}{\partial q_i} \\ &\quad + A_1 \frac{\partial \hat{u}_1}{\partial q_i} + A_2 \frac{\partial \hat{u}_2}{\partial q_i} + A_3 \frac{\partial \hat{u}_3}{\partial q_i} \end{aligned} \tag{1.3-18}$$

where the first three terms on the right-hand side involve partial derivatives of the scalar components $A_1(q_1, q_2, q_3)$, $A_2(q_1, q_2, q_3)$, and $A_3(q_1, q_2, q_3)$ of the vector field, each in their respective unit-vector direction. These terms are therefore handled as in Eqs. (1.3-1), (1.3-3), and (1.3-5). The last three terms involve coordinate derivatives of unit vectors and must be considered—a point entirely missed when the Cartesian system is used.

The vital difference is that spatial derivatives of unit vectors in Cartesian coordinates are all zero, but that derivatives of unit vectors with respect to coordinates that are curved in space often are not. One might think that the derivative of a vector whose length is constant has to be zero. However, this is not the case.

In general it can be shown[11] that

$$\frac{\partial \hat{u}_i}{\partial q_i} = -\frac{\hat{u}_j}{h_j}\frac{\partial h_i}{\partial q_j} - \frac{\hat{u}_k}{h_k}\frac{\partial h_i}{\partial q_k} \tag{1.3-19}$$

and

$$\frac{\partial \hat{u}_i}{\partial q_j} = \frac{\hat{u}_j}{h_i}\frac{\partial h_j}{\partial q_i} \tag{1.3-20}$$

where $i = 1,2,3;\ j = 2,3,1$ and $k = 3,1,2$, in that order. Further, if the derivative of a unit vector is not zero, it will always be at right angles to that unit vector. Thus,

$$\hat{u}_i \cdot \frac{\partial \hat{u}_i}{\partial q_j} = 0 \tag{1.3-21}$$

*Example: The movement of a clock hand to illustrate the need for coordinate derivative of a unit vector.*

Think of the hand of a clock. In a cylindrical coordinate system (or just a polar coordinate system because the problem is just 2D), our coordinates $q_1$ and $q_2$ are $r$ and $\phi$, and the metric coefficients $h_1$ and $h_2$ are $1$ and $r$, respectively. Representing the clock hand as $\hat{u}_r$, the $\phi - coordinate$ partial derivative of $\hat{u}_r$ can be found from (1.3-20):

$$\begin{aligned} \frac{\partial \hat{u}_r}{\partial q_\phi} &= \frac{\hat{u}_\phi}{h_r}\frac{\partial h_\phi}{\partial r} = \frac{\hat{u}_\phi}{1}\frac{\partial r}{\partial r} \\ \frac{\partial \hat{u}_r}{\partial \phi} &= \hat{u}_\phi \end{aligned} \tag{1.3-22}$$

Therefore, the rate of change of the clock hand, represented by the unit vector in the $r$-direction with respect to the azimuthal $\phi$ direction is equal to the unit vector in the $\phi$ direction. Further, by applying this result to (1.3-21), we see that

$$\hat{u}_r \cdot \frac{\partial \hat{u}_r}{\partial \phi} = \hat{u}_r \cdot \hat{u}_\phi = 0 \tag{1.3-23}$$

Thus, the derivative of the unit vector that is always pointing in the direction of the clock hand, i.e. $\hat{u}_r$, with respect to $\phi$ is at right angles to $\hat{u}_r$ and, in fact, is in the $\hat{u}_\phi$ direction. This orthogonal result will always be so because the unit vector being differentiated does not change length.

## References

1. Douglas H. Werner and Raj Mittra, *Frontiers in Electromagnetics*, IEEE Press, Piscataway, NJ (2000).
2. Max Born and Emil Wolf, *Principles of Optics*, 6th ed., Cambridge University Press (1980).
3. Govind P. Agrawal, *Nonlinear Fiber Optics*, 3rd Ed., Academic Press, London (2001), pp. 17ff.
4. Clifford R. Pollock, *Fundamentals of Optoelectronics*, McGraw-Hill, New York (1994), pp. 190,200, e.g.
5. Robert W. Boyd, *Nonlinear Optics*, Academic Press, London (1991).
6. Paul N. Butcher and David Cotter, *The Elements of Nonlinear Optics*, Cambridge University Press (1991), p. 4ff.
7. Eugenio Iannone, Francesco Matera, Antonio Mecozzi and Marina Settembre, *Nonlinear Optical Communication Networks*, Wiley, New York (1998), p. 36ff.
8. William H. Hayt, Jr. and John A. Buck, *Engineering Electromagnetics*, 6th ed., McGraw-Hill, New York, (2001).
9. Matthew N. O. Sadiku, *Elements of Electromagnetics*, 3rd ed., Oxford University Press, (2000).
10. Clayton R. Paul, *Electromagnetics for Engineers: With Applications to Digital Systems and Electromagnetic Interference*, Wiley, New York (2003).
11. A. I. Lure, *Three-Dimensional Problems of the Theory of Elasticity*, D. B. McVean, Trans.,Wiley Interscience, Hoboken, NJ (1964).

## Chapter 2
# Vector Algebra Review

The purpose of this chapter is to review some of the salient operations involving scalar and vector fields and to broaden these concepts to dyadics and tensors in general. Here we briefly discuss variant and invariant scalars, the concept of scalar and vector fields, and the utility of phasor forms of these quantities. Classical arithmetic vector operations of addition, subtraction, and dot and cross products are discussed along with physical applications of these. The direct vector-vector product is mentioned in Section 2.4.3 as having a dyadic resultant; however, the details of this process are left to later chapters.

The basic building blocks of open and closed line and surface integrals of vector fields are discussed. These are essential for both the definitions of vector differential operators, covered in Chapter 4, and the integral forms that shape the basis of divergence, Stokes', and Green's theorems covered in Chapter 5. Other highly useful applications of dot- and cross-product operations conclude the sections of this chapter. These are vector field direction lines and equivalue surfaces of scalar fields.

## 2.1 Variant and Invariant Scalars

A quantity is said to be a *scalar* if it has only magnitude, that is, no inherent direction. Quantities such as time, mass, distance, temperature, entropy, energy, electric potential, and pressure have a value at every position in space but lack directionality. These are scalars. Because such quantities are independent of the orientation of a coordinate system, they are called *invariant scalars*. Coordinates of a point and components of a vector are also scalars; however, these quantities change with coordinate displacements and rotations and therefore are *variant scalars.*

## 2.2 Scalar Fields

In general, scalar fields are quantities that can be represented by functions of space and time. For example, a quantity might be described as a function of four independent variables, such as three orthogonal coordinates $q_1, q_2, q_3$ and time $t$. That is, at every point in space described uniquely by $q_1, q_2, q_3$ and at each instant of time described by $t$, a field, such as electric potential (frequently written as

$V$ or $\Phi$) can be described by a single-valued scalar function of these four independent variables as

$$V(q_1, q_2, q_3, t) = V(\bar{r}, t) \quad \text{volts} \tag{2.2-1a}$$

or alternatively,[1]

$$\Phi(q_1, q_2, q_3, t) = \Phi(\bar{r}, t) \quad \text{volts} \tag{2.2-1b}$$

The right-hand side of these expressions incorporates $\bar{r}$*-space notation* described in Section 1.1.4—a shorthand notation where the coordinates $q_1, q_2, q_3$ are represented by $\bar{r}$—which denotes a point in space with or without regard to a coordinate system as shown in Fig. 1.1-1. For Cartesian, cylindrical or spherical coordinates, the arguments $x, y, z$ or $r, \phi, z$ or $r, \theta, \phi$, respectively, are replaced by $\bar{r}$, for example.

Thus, electric potential is a *scalar field*—or more specifically, an *electric potential field* in space and time. $V$ (or $\Phi$) is an invariant scalar since its value at a specific point in space and at a specific time is independent of any chosen coordinate system and independent of any rotation or displacement of a coordinate system. See Fig. 2.2-1(a).

## 2.3 Vector Fields

Quantities such as force, velocity, displacement, electric and magnetic field intensities, and electric and magnetic flux densities are vectors because each of these has a magnitude and a *single* direction.* The magnitude and direction of each of these can vary in space and time. In 3D space, vectors can be described by three scalar magnitudes that represent components in some orthogonal coordinate system. Although a vector may be invariant to a coordinate transformation, its component magnitudes are, in general, not. Since each component is a scalar field that varies in space and time as in Eqs. [2.2-1(a)–(b)], the entire vector is a function of space and time.

---

* Other quantities have magnitude and direction, which are not vectors. For example, stress has *dual directional compoundedness*—not a "single" direction as emphasized in this definition. Such quantities are dyadics, not vectors. Dyadics and other tensors are discussed in greater detail in Chapters 3 and 4.

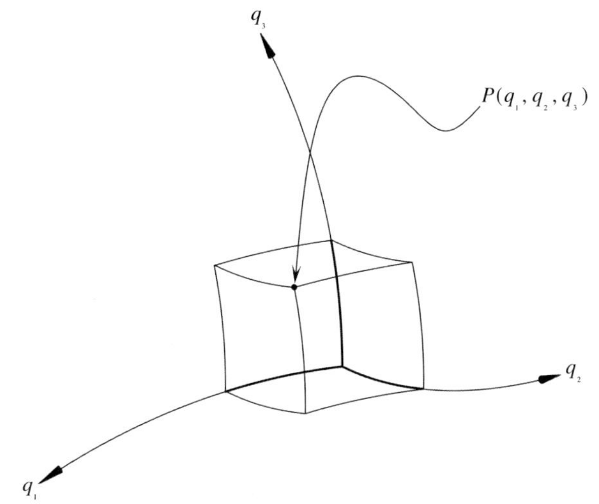


(a) Scalar field $V(q_1, q_2, q_3, t)$ at the coordinate point in space $P(q_1, q_2, q_3)$ and at time *t*.

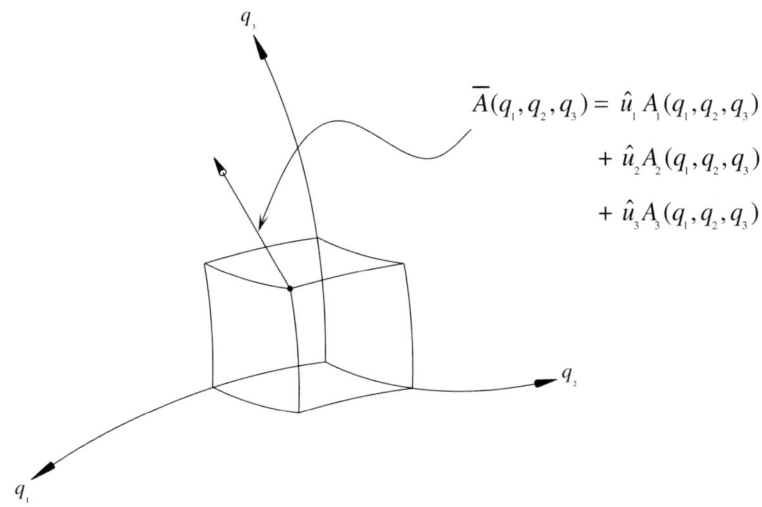


(b) Vector field $\overline{A}(q_1, q_2, q_3, t)$ at the coordinate point in space $P(q_1, q_2, q_3)$ and at time *t*.

**Figure 2.2-1** The concept of (a) scalar and (b) vector fields.

Thus, the vector function of spatial coordinates $q_1, q_2, q_3$ can be thought of as a *vector field.* As with scalars, a vector field such as electric field intensity $\overline{E}$ can be described as a function of space and time as

$$\overline{E}(q_1, q_2, q_3, t) = \overline{E}(\overline{r}, t) \qquad \text{volts/m} \qquad (2.3\text{-}1)$$

where the right-hand side of (2.3-1) uses the same shorthand $\overline{r}$-*space* notation as described earlier for scalar fields. In electromagnetics, the symbol $\overline{E}$ is used for the electric field intensity. Thus, the electric field intensity is a vector field in space and time.

In our 3D space $\overline{E}$ is customarily expanded into three orthogonal components

$$\overline{E} = \hat{u}_1 E_1 + \hat{u}_2 E_2 + \hat{u}_3 E_3 \qquad (2.3\text{-}2)$$

where each component $\overline{E}$ is a scalar function of our four independent variables

$$E_i(q_1, q_2, q_3, t) \qquad (2.3\text{-}3)$$

where $i$ = 1, 2, 3. Figure 2.2-1(b) illustrates a vector field $\overline{A}$ at the point $q_1, q_2, q_3$ at time $t$. Notice that the vector direction is entirely independent of the position in space.

## 2.4 Arithmetic Vector Operations

This section briefly deals with arithmetic analyses involving vector addition, subtraction, dot and cross products, and certain allowable division operations. Some of the more elementary applications of vector addition and subtraction can be found in Appendix A.

### 2.4.1 Commutative and associative laws in vector addition and subtraction

Consider two vectors $\overline{A}$ and $\overline{B}$ at a point in space. When added, we find (in Appendix A) that the sum of vectors is independent of the order in which they are added. Thus, vector addition obeys the *commutative law*:

$$\overline{A} + \overline{B} = \overline{B} + \overline{A} \qquad (2.4\text{-}1)$$

When a third vector $\overline{C}$ is added to $\overline{A}+\overline{B}$, the resultant vector $\overline{H}=(\overline{A}+\overline{B})+\overline{C}$. As is shown graphically in Appendix A, if $\overline{A}$ is added to $\overline{B}+\overline{C}$, the result is the same vector $\overline{H}=\overline{A}+(\overline{B}+\overline{C})$. That is, regardless of whether $\overline{B}$ is first associated with $\overline{A}$ or first associated with $\overline{C}$, the same resultant vector $\overline{H}$ is obtained. This feature of vector addition is referred to as the *associative law*:

$$(\overline{A}+\overline{B})+\overline{C}=\overline{A}+(\overline{B}+\overline{C}) \tag{2.4-2}$$

Alternatively, if $\overline{C}$ is added to $\overline{A}$ first and then $\overline{B}$ is added to the result, we obtain the same vector $\overline{H}$.

Subtraction follows these laws as well. Consider $\overline{A}-\overline{B}=\overline{D}$. In order to test commutative and associative laws, let us alternatively consider $-\overline{B}+\overline{A}$. As graphically shown in Appendix A, this latter commutation of vectors yields the same vector $\overline{D}$. Thus, subtraction of vectors is both commutative

$$\overline{A}-\overline{B}=-B+A \tag{2.4-3}$$

and associative

$$(\overline{A}-\overline{B})+\overline{C}=\overline{A}+(-\overline{B}+\overline{C}) \tag{2.4-4}$$

These conclusions are to be expected, since subtraction is a special case of addition.

### 2.4.2 Multiplication or division of a vector by a scalar

When a vector is multiplied by a scalar, the resultant vector is in the same direction but its magnitude is simply the product of the magnitude of the original vector multiplied by the scalar. The vector $\overline{a}$ when multiplied by the scalar $m$ yields another vector $\overline{F}$ whose magnitude is the product of $m$ and $|\overline{a}|$ and whose direction is that of $\overline{a}$. Thus, in the case of force-mass-acceleration relation,

$$\overline{F} = m\overline{a} \tag{2.4-5}$$

notice that the magnitude of the resultant vector will depend upon that of the scalar. For example, if $\overline{a}$ is acceleration (m/sec$^2$) and $m$ is mass (kg), the resultant is force $\overline{F}$ (kg m/s$^2$ or newtons). To illustrate, if the acceleration is 4 m/s$^2$ in the $\hat{u}_x$ direction and $m$ is 2.25 kg, then $\overline{F} = 9\hat{u}_x$ newtons. Note that the acceleration and force are in the same direction, that of the unit vector $\hat{u}_x$. We will see later that when the vectors on the right and left are differently directed, the multiplicative factor cannot be a simple scalar [such as in Eq. (2.4-9) below].

The distributative laws of algebra also apply to scalar-vector products. Namely

$$n(\overline{A}+\overline{B}+\overline{C}) = n\overline{A}+n\overline{B}+n\overline{C} \tag{2.4-6}$$

and

$$(n+m)\overline{A} = n\overline{A}+m\overline{A} \tag{2.4-7}$$

It goes without saying that when a vector is divided by a scalar, the expression $\overline{D} = \varepsilon\overline{E}$ can just as well be written as

$$\overline{E} = \frac{\overline{D}}{\varepsilon} \tag{2.4-8}$$

In the latter expression, each of the three components of $\overline{D}$ are simply divided by $\varepsilon$ to obtain the three components of $\overline{E}$. Thus, in Eq. (2.4-8) $\overline{E}$ and $\overline{D}$ are everywhere parallel.

However, we cannot simply divide by a vector or any other tensor [unless, of course, the tensor is a scalar as in Eq. (2.4-8)]. For example, in electrostatics or electromagnetics, the flux density vector field $\overline{D}$ induced in an anisotropic dielectric by an applied electric field intensity $\overline{E}$ is given by the constitutive relation

$$\overline{D} = \overline{\overline{\varepsilon}}\cdot\overline{E} \tag{2.4-9}$$

where $\overline{\overline{\varepsilon}}$ is a dyadic. The dot product operation in Eq. (2.4-9)—namely the dot product of a dyadic with a vector—is carried out in detail in the next Chapter (Section 3.4). For now, suffice it to say that $\overline{E}$ and $\overline{D}$ are not necessarily everywhere parallel for the case of anisotropic media. The divide-by operation for the case of a dyadic requires that the inverse of $\overline{\overline{\varepsilon}}$ be first taken and then $\overline{E}$ determined by

$$\overline{E} = \left(\overline{\overline{\varepsilon}}\right)^{-1}\cdot\overline{D} \tag{2.4-10}$$

where the inverse $(\bar{\bar{\varepsilon}})^{-1}$ is performed the same as the inverse of a 3 × 3 matrix, [except that the unit dyads in $\bar{\bar{\varepsilon}}$ are transposed in $(\bar{\bar{\varepsilon}})^{-1}$]. See Sections 3.3 and 3.4.

### 2.4.3 Vector-vector products

We next look at vector-vector products in orthogonal systems. There are three fundamentally distinct ways to perform product operations between two vectors. Each has an entirely different result. The first is the *vector dot product*. The second is the *vector cross product*, and the third we will call a *direct product.** Since the vector-vector dot product always yields a directionless scalar, it is also called a *scalar product*. Similarly, since the vector-vector cross product always yields a vector, it is also referred to as a *vector product*. The "direct product" of two vectors yields a *dyadic*, which is described in Chapter 3 and applied in Chapter 4. Cantrell[2] and many other cutting edge references on tensor calculus include this latter product—first between vectors and then involving tensors in general.

#### 2.4.3(a) Restricted use of the terms "scalar product" and "vector product"

Part (b) below equates the operations "dot product" with "scalar product" and "inner product." Also, Part (c) equates "cross product" with "vector product," "external product," and "outer product." However, it should be emphasized that the expressions "scalar product" and "vector product" are suitable only for the case of the dot or cross products being taken between quantities that are both vectors. If either or both of the quantities are tensors having a directional compoundedness greater than one, then the dot product no longer yields a scalar and the cross product no longer yields a vector. Further, dot- and cross-product operations are both undefined and unnecessary when scalars are involved.

---

* The term *"direct product"* is used here to mean that the operation is conducted without a dot- or cross-product type of a process. That is, each component of the first vector individually multiplies each component of the second vector. Thus, there are nine such multiplication operations each with two directions inherent within their respective operation—one associated with the first vector and the other associated with the second. The resultant of a "direct product" between two vectors is not a scalar nor a vector because of this dual-directional nature. This is a *dyadic*. For those already familiar with tensors, the term "direct product" is used here synonymously with *"tensor product,"* which uses the symbol ⊗ to denote a product between all combinations of the components each vector (in our case) or between two tensors, in general, as we shall see later.

*Therefore, the expressions "scalar product" and "vector product" should be dropped from our nomenclature when scalars, dyadics, triadics, etc. are involved. They are accurate expressions only when both quantities are vectors and should be avoided for all other quantities. In general the terms "inner product" and "dot product" may be interchanged*[3,6,7]*. "Exterior product", "outer product" and "cross product" are occasionally used interchangeably*[13,14]*; however, in broader contexts these terms are distinct*[2,13,14]*.*

### 2.4.3(b) Dot product and the Kronecker delta

The dot product between two vectors $\overline{A}$ and $\overline{B}$ is spoken as "$\overline{A}$ dot $\overline{B}$," written as $\overline{A} \cdot \overline{B}$, and in elementary texts on vector analysis is defined by[3–5]

$$\overline{A} \cdot \overline{B} = |\overline{A}||\overline{B}|\cos\theta_{AB} = \overline{B} \cdot \overline{A} \tag{2.4-11}$$

where $\theta_{AB}$ is the angle between $\overline{A}$ and $\overline{B}$. Notice that if the two vectors are at right angles, $\theta_{AB} = 90^\circ$, and the dot product equals zero. As stated in the previous paragraph, the dot product between two vectors is also known as the scalar product, because the resultant is a scalar.

Another name for this operator is *inner product*, a term used historically in mathematical parlance between real or complex vectors,[6–10] especially when the operation deals with tensors in general. Furthermore, we will find that there can be multiple inner-product operations, like *double-dot product* and *triple-dot product* operations when dealing with tensors.[11,12] This will be discussed in Chapter 3, where more general concepts of vector analysis are applied in the context of tensors.

In any generalized orthogonal right-hand coordinate system having coordinates $q_1, q_2, q_3$, where the right-hand rule applies in the order 1, 2, 3, 1, 2, unit vectors in each of the three directions $\hat{u}_1$, $\hat{u}_2$, $\hat{u}_3$ have well-determined dot product relationships. The nine combinations of dot product operations can all be accounted for with the convenient *Kronecker delta* $\delta_{ij}$ notation[13] (valid for orthogonal systems) by

$$\hat{u}_i \cdot \hat{u}_j = \delta_{ij} = \begin{cases} 1 & \text{for } i=j \\ 0 & \text{for } i \neq j \end{cases} \tag{2.4-12}$$

where $i = 1,2,3$ and $j = 1,2,3$.

Let us illustrate the inner-product nature of the dot-product operation taken between two vectors. Let $\overline{A}=\hat{u}_1A_1+\hat{u}_2A_2+\hat{u}_3A_3$ and $\overline{B}=\hat{u}_1B_1+\hat{u}_2B_2+\hat{u}_3B_3$, where $A_1, A_2, A_3,$ and $B_1, B_2, B_3$ are the scalar components of the two vectors, respectively. These can be written $\overline{A}=\hat{u}_iA_i$ and $\overline{B}=\hat{u}_jB_j$ in shortened tensor notation, such as in Eq. (1.2-3), where the summation from one to three is suppressed. The dot product of $\overline{A}$ and $\overline{B}$ when expanded in explicit standard notation becomes

$$\begin{aligned}\overline{A}\cdot\overline{B}= &\quad \hat{u}_1\cdot\hat{u}_1\, A_1B_1 + \hat{u}_1\cdot\hat{u}_2\, A_1B_2 + \hat{u}_1\cdot\hat{u}_3\, A_1B_3\\ &+ \hat{u}_2\cdot\hat{u}_1\, A_2B_1 + \hat{u}_2\cdot\hat{u}_2\, A_2B_2 + \hat{u}_2\cdot\hat{u}_3\, A_2B_3\\ &+ \hat{u}_3\cdot\hat{u}_1\, A_3B_1 + \hat{u}_3\cdot\hat{u}_2\, A_3B_2 + \hat{u}_3\cdot\hat{u}_3\, A_3B_3\end{aligned} \qquad (2.4\text{-}13)$$

Applying Eq. (2.4-12) results in six of the above nine terms going to zero, namely the off-diagonal terms, and the diagonal unit vector dot products equaling unity. Therefore, only the diagonal terms of Eq. (2.4-13) survive, and $\overline{A}\cdot\overline{B}$ becomes

$$\begin{aligned}\overline{A}\cdot\overline{B}&=A_1B_1+A_2B_2+A_3B_3\\ &=\sum_{i=1}^{3}A_iB_i\end{aligned} \qquad (2.4\text{-}14)$$

which is, of course, a scalar.

In tensor notation, $\overline{A}\cdot\overline{B}$ is written as

$$(\hat{u}_iA_i)\cdot(\hat{u}_jB_j)=A_iB_j\delta_{ij}=A_iB_i \qquad (2.4\text{-}15)$$

where the six off-diagonal zeros and the three diagonal ones from the Kronecker delta of Eq. (2.4-12) applied to Eq. (2.4-13) and the summation in Eq. (2.4-14) are implied. In fact, more fundamental than Eq. (2.4-11) for the definition of the inner-product operation between two vectors is the definition given in Eq. (2.4-15), because this latter form is a special case of a dot product between tensors in general, as we shall see in Section 3.4. First, we show two examples of the use of the dot product—one from vector algebra and the other from vector calculus.

*Application from vector algebra: Projection of one vector onto another.*

The projection of one vector onto another can readily be obtained with the application of the dot product. Let us define a unit vector $\hat{u}_A$ in the direction of $\overline{A}$ as

$$\hat{u}_A = \frac{\overline{A}}{|\overline{A}|} \tag{2.4-16}$$

The projection of a second vector $\overline{B}$ onto a line containing $\overline{A}$ is the scalar component of $\overline{B}$ in the direction of $\overline{A}$. Thus, the scalar projection of $\overline{B}$ onto $\overline{A}$ is

$$B_A = |\overline{B}|\cos\theta_{AB} \tag{2.4-17}$$

as shown in Fig. 2.4-1.

Multiplying by $|\overline{A}|/|\overline{A}|$ does not change the result:

$$B_A = \frac{|A||B|\cos\ \theta_{AB}}{|\overline{A}|}$$

Since the numerator is the dot product $\overline{A}\cdot\overline{B}$ by definition, we may write the scalar projection of $\overline{B}$ onto vector $\overline{A}$ as

$$B_A = \frac{\overline{A}\cdot\overline{B}}{|\overline{A}|} = \overline{B}\cdot\hat{u}_A = \hat{u}_A\cdot\overline{B} \tag{2.4-18}$$

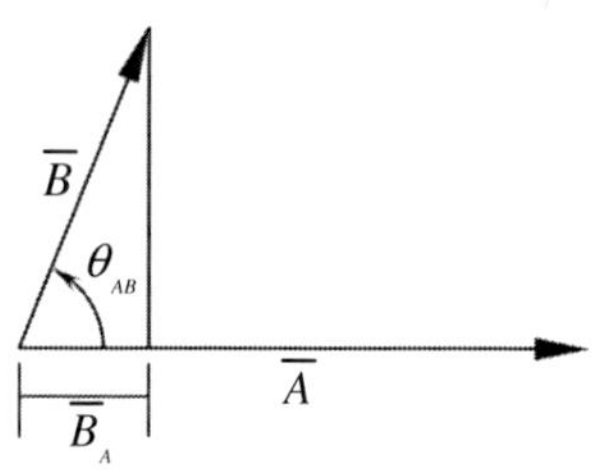


**Figure 2.4.1** Scalar projection of vector $\overline{B}$ onto vector $\overline{A}$.

*The projection or component of a vector onto a line containing another vector can be determined by taking the dot product of the first vector with the unit vector in the direction of the second. The result is a scalar.*

Furthermore, the *vector* projection of $\overline{B}_A$ onto the line containing $\overline{A}$ may be shown simply by including the unit vector in the direction of $\overline{A}$ in the expression for the scalar projection. The vector projection of $\overline{B}_A$ onto $\overline{A}$ is then

$$\overline{B}_A = \hat{u}_A B_A = \hat{u}_A(\overline{B} \cdot \hat{u}_A) \qquad (2.4\text{-}19)$$

Other applications of the dot product are provided in Appendix A.

*Applications from vector calculus: Dot products in line and surface integrands.*

Other important applications of the dot product deal with line and surface integrals. It is frequently important to integrate a vector field $\overline{F}$ along a path defined by $\overline{d\ell}$ or over a surface $\overline{da}$. The vector field $\overline{F}$ may, of course, have any directional orientation, not necessarily in the same direction as $\overline{d\ell}$ or in the same direction as the surface normal $\overline{da}$. Whenever the component of the field along the path or normal to the surface needs to be summed differentially—that is, integrated—the dot product is used.

Let us first take the case of the line integral. In many applications, we wish to know the potential of a vector field between two points in space along a given path (such as work in the case of a force field or electric potential in the case of an electric field). In this case, we are looking for the sum of all tangential components of the vector field along all differential elements of length along the given path. This is the line integral. Thus, we apply the dot product of the vector field and the vector differential lengths in order to accumulate just the tangential components along the path. By integrating, we then obtain the potential of the field from a point *a* to a point *b*. For any vector field $\overline{F}$ this is obtained by

$$\int_a^b \overline{F} \cdot \overline{d\ell} \qquad (2.4\text{-}20)$$

which is referred to as an *open line integral.* This is an example of the use of the dot product in vector calculus. Before providing other such applications, the utility of an open line integral is illustrated by the following example.

*Example: Potential energy and electric potential.*

If $\overline{F}$ is a force field (in newtons), Eq. (2.4-20) represents joules of potential energy—a scalar—between points *a* and *b*. If $\overline{F}$ is electric field intensity (in newtons per coulomb), Eq. (2.4-20) is volts of electric potential—also a scalar—between points *a* and *b*. More will be said about this important integral in Section 5.1. Two numerical examples are given in Subsection 5.1.1. The first is for a path-independent case. The second is for a path-dependent case.

When the line integral is closed to form a loop, Eq. (2.4-20) takes the form of a *closed line integral*

$$\oint \overline{F} \cdot \overline{d\ell} \tag{2.4-21}$$

This integral is called the *circulation* of the vector field $\overline{F}$ and is written as

$$\mathrm{circ}(\overline{F}) = \oint \overline{F} \cdot \overline{d\ell} \tag{2.4-22}$$

If the circulation is zero, the vector field $\overline{F}$ is said to be *conservative* or *irrotational*. When the circulation is not equal to zero, the vector field $\overline{F}$ is said to be *solenoidal* or *rotational*.

*Example: Ampere's circuital law.*

An example of the utility of Eqs. (2.4-21) and (2.4-22) is Ampere's circuital law, which states that the integration of the tangential component of the magnetic field intensity $\overline{H}$ around any closed path is equal to the total electric current enclosed by that path. This is described in Section 5.4, especially by Eq. 5.4-1.

Two other common applications of the use of the dot product are in open and closed surface integrals. Again, the vector field $\overline{F}$ may have any directional orientation, not necessarily in the same direction as the surface normal $\overline{da}$. Whenever the component of the field normal to the surface needs to be summed differentially (integrated), the dot product is used. This is the case, for example, when the vector field $\overline{F}$ is a flux density and the total flux is desired. The flux of the vector field $\overline{F}$ through an area *A* is determined by integrating the dot product of the flux density over every differential vector element of area $\overline{da}$ defined by Eq. (1.2-5) and Fig. 1.2-2:

$$\int_A \overline{F} \cdot \overline{da} \tag{2.4-23}$$

where the area $A$ is bounded (by a closed line).

When the surface integral is closed to enclose a volume, Eq. (2.4-23) takes the form

$$\oint \overline{F} \cdot \overline{da} \tag{2.4-24}$$

These integrals with dot products in the integrand are frequently used in disciplines of mathematical physics, such as quantum physics and electromagnetics. The dot product in the integrand is simply a convenient way to sum only the component of $\overline{F}$ at each differential element of surface over which the integration takes place that lies normal to that surface element.

Examples of Eqs. (2.4-23) and (2.4-24) can be found in Sections 5.2.2, 5.3.1, and 5.3.4.

### 2.4.3(c) Cross product and the Levi-Civita symbol

The "cross" product of vector $\overline{A}$ with another vector $\overline{B}$ is spoken as "$\overline{A}$ cross $\overline{B}$" and written as $\overline{A} \times \overline{B}$. The cross product is defined by

$$\overline{A} \times \overline{B} = \hat{u}_{\perp_{A \to B}} |A||B| \sin\theta_{\mathrm{AB}} \tag{2.4-25}$$

where $\hat{u}_{\perp_{A \to B}}$ is a unit vector normal to the plane containing $\overline{B}$ and $\overline{A}$ and is in a direction given in a right-hand sense—namely by aligning the fingers of your right hand along the direction of $\overline{A}$ and turning them into the direction of $\overline{B}$ so that your thumb points in the direction of $\hat{u}_{\perp_{A \to B}}$. The angle $\theta_{AB}$ is the angle made in so doing.

#### (i) Commutative and distributive laws for cross products

From Eq. (2.4-25), note that $\overline{A} \times \overline{B} = -\overline{B} \times \overline{A}$. Thus, the commutative law does not hold for the cross product operation. However, the *distributive law* does hold. Namely,

$$
\begin{aligned}
&(\bar{A}+\bar{B}+\cdots)\times(\bar{M}+\bar{N}+\cdots)\\
&\quad=\bar{A}\times\bar{M}+\bar{A}\times\bar{N}+\cdots\\
&\quad+\bar{B}\times\bar{M}+\bar{B}\times\bar{N}+\cdots\\
&\qquad+\cdots
\end{aligned}
\tag{2.4-26}
$$

(ii) Vector cross products and the Levi-Civita symbol

Unit vectors in each of three orthogonal directions $\hat{u}_1$, $\hat{u}_2$, $\hat{u}_3$ have well-determined cross-product relationships. These relationships are described conventionally in the following paragraph and described with the elegance of the *Levi-Civita* symbol in the subsequent paragraph.

The cross product of unit vectors in 3D space becomes trivalued, namely, $-1$, 0, and +1. The usual process used in sophomore-level texts to explain this trivalued system is to first point out that $\hat{u}_i \times \hat{u}_i = 0$ because $\theta_{ij} = 0$ and the $\sin\theta_{ij} = 0$ in Eq. (2.4-25). Further, $\hat{u}_i \times \hat{u}_{i+1} = +\hat{u}_{i+2}$ where $i$ = 1, 2, 3; $i$ + 1 = 2, 3, 1 and $i$ + 2 = 3, 1, 2, because $\theta_{(i)(i+1)} = \pi/2$ and $\sin\theta_{(i)(i+1)} = 1$. The right-hand rule specifies that direction 1 crossed into direction 2 yields positive direction 3, or direction 2 crossed into direction 3 yields positive direction 1, and direction 3 crossed into direction 1 yields positive direction 2. However, $\hat{u}_i \times \hat{u}_{i+2} = -\hat{u}_{i+1}$, where $i$ = 1, 2, 3; $i$ + 2 = 3, 1, 2; and $i$ + 1 = 2, 3, 1. The minus sign is needed because when direction 1 is crossed into direction 3 the thumb points opposite to (or the negative of) direction 2. Likewise, 2 into 1 yields the negative of direction 3 and 3 into 2 yields the negative of direction 1. The angle from 1 to 3 may be taken as $-\pi/2$ since the angle from 3 to 1 is $\pi/2$. Thus, $\theta_{(i)(i+2)} = -\pi/2$ and $\sin\theta_{(i)(i+2)} = -1$.

However, the *Levi-Civita* symbol $\in_{ijk}$ shortcuts the discussion in the preceding paragraph. If one calls the sequence 1,2,3,1,2 *cyclic*, the sequence 3,2,1,3,2 *acyclic*, and cases where any two adjacent indices are the same *noncyclic*, we define the Levi-Civita symbol as[14]

$$
\in_{ijk} = \begin{cases} 1 & \textit{cyclic} \\ 0 & \textit{noncyclic} \\ -1 & \textit{acyclic} \end{cases}
\tag{2.4-27}
$$

and therefore,

$$
u_i \times u_j = \in_{ijk} u_k
\tag{2.4-28}
$$

which is a tensor notation formulation with the unit vector hats implied.

The cross product of our vector $\overline{A}$ with $\overline{B}$ in tensor notation can then be defined as

$$A_i u_i \times B_j u_j = A_i B_j \in_{ijk} u_k \tag{2.4-29}$$

(iii) Area formulas using cross products

In Section 1.2 differential area was defined and discussed without the benefit of the cross product. A description of the vector differential area [Eq. (1.2-5)] can now be expressed as

$$\overline{da} = \hat{u}_k \left|\overline{d\ell_i}\right|\left|\overline{d\ell_j}\right| = \frac{\overline{d\ell_i} \times \overline{d\ell_j}}{\left|\overline{d\ell_i} \times \overline{d\ell_j}\right|}\left|\overline{d\ell_i}\right|\left|\overline{d\ell_j}\right| \tag{2.4-30}$$

or more simply in tensor notation as

$$\overline{da} = d\ell_i u_i \times d\ell_j u_j = d\ell_i d\ell_j \in_{ijk} u_k \tag{2.4-31}$$

Note also that the area of the parallelogram with adjacent sides $\overline{A}$ and $\overline{B}$ is the magnitude of the cross product where

$$\text{Area} = |A||B| \sin\theta_{AB} = |\overline{A} \times \overline{B}| \tag{2.4-32}$$

This is illustrated in Fig. 2.4-2.

Other applications of the cross product include finding the moment of a force acting at a distance, finding the force on a current-carrying conductor in a magnetic field, and dealing with the mechanics of gyroscopes, among many others.

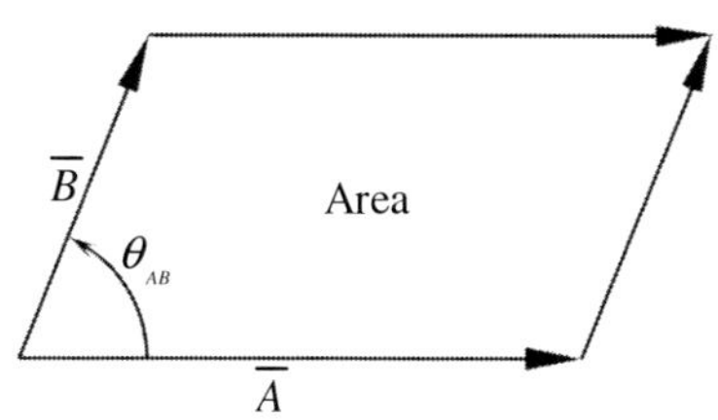


**Figure 2.4-2** The area of a parallelogram as $\left|\overline{A} \times \overline{B}\right|$.

## (iv) Cross product coordinate expansion

Using the same vectors $\overline{A} = \hat{u}_i A_i$ and $\overline{B} = \hat{u}_j B_j$ as before, but using tensor notation, the cross product takes the form

$$\begin{aligned}
\overline{A}\times\overline{B} = {} & \overset{0}{\cancel{\hat{u}_1 \times \hat{u}_1}}\ A_1B_1 + \overset{\hat{u}_3}{(\hat{u}_1 \times \hat{u}_2)}\ A_1B_2 + \overset{-\hat{u}_2}{(\hat{u}_1 \times \hat{u}_3)}\ A_1B_3 \\
& + \overset{-\hat{u}_3}{(\hat{u}_2 \times \hat{u}_1)}\ A_2B_1 + \overset{0}{\cancel{\hat{u}_2 \times \hat{u}_2}}\ A_2B_2 + \overset{\hat{u}_1}{(\hat{u}_2 \times \hat{u}_3)}\ A_2B_3 \\
& + \overset{\hat{u}_2}{(\hat{u}_3 \times \hat{u}_1)}\ A_3B_1 + \overset{-\hat{u}_1}{(\hat{u}_3 \times \hat{u}_2)}\ A_3B_2 + \overset{0}{\cancel{\hat{u}_3 \times \hat{u}_3}}\ A_3B_3
\end{aligned} \tag{2.4-33}$$

From Eq. (2.4-28), the cross product factors become 0, $\hat{u}_k$, or $-\hat{u}_k$, where $k = 1, 2, 3,$ as shown in Eq. (2.4-33).

Collecting terms in each of the three coordinate directions,

$$\begin{aligned}
\overline{A}\times\overline{B} = {} & \hat{u}_1(A_2B_3 - A_3B_2) \\
& + \hat{u}_2(A_3B_1 - A_1B_3) \\
& + \hat{u}_3(A_1B_2 - A_2B_1)
\end{aligned} \tag{2.4-34}$$

Notice that this can also be represented in determinate form as

$$\overline{A}\times\overline{B} = \begin{vmatrix} \hat{u}_1 & \hat{u}_2 & \hat{u}_3 \\ A_1 & A_2 & A_3 \\ B_1 & B_2 & B_3 \end{vmatrix} \tag{2.4-35}$$

Alternatively, tensor notation can be used in conjunction with the Levi-Civita symbol to express $\overline{A}\times\overline{B}$ as

$$(\hat{u}_i A_i)\times(\hat{u}_j B_j) = \hat{u}_k \in_{ijk} A_i B_j \tag{2.4-36}$$

in its ultimate beauty and simplicity, but still preserving all of the operations of Eq. (2.4-33) resulting in the six nonzero terms of (2.4-34), including the three minus signs.

Whereas Eqs. (2.4-15) and (2.4-33) have an elegance in their exquisite simplicity when applied to vector operations, their real power and attractiveness comes in dealing with inner and external products of tensors.

### 2.4.3(d) Triple vector products

Two identities involving vector-vector-vector products are useful in vector algebra problems. These are

$$(\overline{A}\times\overline{B})\cdot\overline{C}=(\overline{B}\times\overline{C})\cdot\overline{A}=(\overline{C}\times\overline{A})\cdot\overline{B} \qquad (2.4\text{-}37)$$

and

$$\overline{A}\times(\overline{B}\times\overline{C})=(\overline{A}\cdot\overline{C})\overline{B}\ -\ (\overline{A}\cdot\overline{B})\overline{C} \qquad (2.4\text{-}38)$$

or alternatively,

$$(\overline{A}\times\overline{B})\times\overline{C}=(\overline{A}\cdot\overline{C})\overline{B}-(\overline{B}\cdot\overline{C})\overline{A} \qquad (2.4\text{-}39)$$

It is left as a problem for the student to derive the above expressions. These can be done by expansion into components as in Eqs. (2.4-14) and (2.4-34). Alternatively, however, after demonstrating these three identities using Eqs. (2.4-14) and (2.4-34), it would be beneficial to the student to repeat these exercises by applying Eqs. (2.4-15) and (2.4-36), and to repeat this process until the true elegance of the latter is felt.

## 2.5 Scalars, Vectors, Dyadics, and Tensors as Phasors

Recall from Section 1.1.4 that whenever fields of any rank vary *time harmonically*, namely as $e^{j\omega t}$, the fields may be written with this time harmonic factor suppressed. We set up the underscore as a notational technique to denote this action. Thus, we have scalar phasors, $\underline{\Phi}(q_1,q_2,q_3)$ and vector phasors, $\underline{\overline{E}}(q_1,q_2,q_3)$. This, of course, can be extended to dyadics, written as

$$\underline{\overline{\overline{A}}}(q_1,q_2,q_3) \qquad (2.5\text{-}1)$$

which is a *dyadic phasor field.* Continuing this generalization, tensors written as phasors also appear with the underscore

$${}_{n_R}\underline{T}(q_1,q_2,q_3) \qquad (2.5\text{-}2)$$

where the presubscript $_{n_R}$ denotes the tensor rank. This is a *tensor phasor field*, where $e^{j\omega t}$ is implied and suppressed as before.

The phasor notation in $\underline{\overline{E}}(q_1,q_2,q_3)$, $\underline{\overline{\overline{A}}}(q_1,q_2,q_3)$, and $_{n_R}\underline{T}(q_1,q_2,q_3)$ implies that each component must be treated as scalar phasors. Therefore, $\underline{\overline{E}}(q_1,q_2,q_3)$ would be expanded as

$$\begin{aligned}\overline{E}(q_1,q_2,q_3,t) &= \hat{u}_1 E_1(q_1,q_2,q_3,t) + \hat{u}_2 E_2(q_1,q_2,q_3,t) + \hat{u}_3 E_3(q_1,q_2,q_3,t) \\ &= \hat{u}_1 \underline{E}_1(q_1,q_2,q_3)e^{j\omega t} + \hat{u}_2 \underline{E}_2(q_1,q_2,q_3)e^{j\omega t} + \hat{u}_3 \underline{E}_3(q_1,q_2,q_3)e^{j\omega t}\end{aligned}$$

or just

$$\underline{\overline{E}}(q_1,q_2,q_3) = \hat{u}_1 \underline{E}_1 + \hat{u}_2 \underline{E}_2 + \hat{u}_3 \underline{E}_3 \qquad (2.5\text{-}3)$$

where

$$\underline{E}_i = \underline{E}_i(q_1,q_2,q_3), \quad i = 1,2,3 \qquad (2.5\text{-}4)$$

are the phasor scalar components of the phasor vector $\underline{\overline{E}}$.

This complex formulation is artificial. It provides a convenient methodology for keeping track of the phases between quantities. In the end, the comparison between the mathematical expectation and measured quantities requires taking only one of the complex parts of the solution. The real part is customarily taken for comparison with measurements.

## 2.6 Vector Field Direction Lines

Suppose we wish to draw a line whose tangent $\overline{dT}$ is everywhere parallel to a known vector field in space. Such a line is referred to as *vector field direction line* or *flow line*. Assuming that the vector field is $\overline{E}$ and is known everywhere in space,

$$\overline{E}(q_1,q_2,q_3) = \hat{u}_1 E_1 + \hat{u}_2 E_2 + \hat{u}_3 E_3 \qquad (2.6\text{-}1)$$

Expanding the unknown differential vector $\overline{dT}$ in the same generalized orthogonal coordinates,

$$\overline{dT} = \hat{u}_1 d\ell_1 + \hat{u}_2 d\ell_2 + \hat{u}_3 d\ell_3 \qquad (2.6\text{-}2)$$

where $q_1$, $q_2$, and $q_3$ are orthogonal coordinates and $d\ell_1, d\ell_2,$ and $d\ell_3$ are differential lengths in each of the coordinate directions, respectively. For example, in spherical coordinates $q_1$, $q_2$, and $q_3$ are

$$q_1 = r \tag{2.6-3a}$$

$$q_2 = \theta \tag{2.6-3b}$$

$$q_3 = \phi \tag{2.6-3c}$$

$$dl_1 = dr \tag{2.6-4a}$$

$$dl_2 = r\ d\theta \tag{2.6-4b}$$

$$dl_3 = r \sin\theta\ d\phi \tag{2.6-4c}$$

In general, the differential lengths may be expressed in terms of differential coordinates by

$$dl_i = h_i dq_i \tag{2.6-5}$$

where the $h_i$ values are referred to as *metric coefficients*, otherwise known as *scale factors*. In the case of spherical coordinates we see from Eqs. (2.6-3a) through (2.6-5) that

$$h_1 = 1 \tag{2.6-6a}$$

$$h_2 = r \tag{2.6-6b}$$

$$h_3 = r\sin\theta \tag{2.6-6c}$$

Substituting Eq. (2.6-5) into Eq. (2.6-1) yields

$$\overline{dT} = \hat{u}_1 h_1 dq_1 + \hat{u}_2 h_2 dq_2 + \hat{u}_3 h_3 dq_3 \tag{2.6-7}$$

Our task is to determine the differential vector $\overline{dT}$ such that it is everywhere tangent to $\overline{E}$. This can be accomplished by noting the following:

*Any two vectors are parallel when their cross product is zero.*

That is, by restricting their cross product to be zero, we have the necessary mathematical construct to find $\overline{dT}$ in terms of the given vector field $\overline{E}$. Thus, if the expression

$$\overline{dT} \times \overline{E} = 0 \tag{2.6-8}$$

is everywhere satisfied, $\overline{dT}$ is everywhere tangent to $\overline{E}$. Equation (2.6-8) may be expanded in matrix form as

$$\begin{vmatrix} \hat{u}_1 & \hat{u}_2 & \hat{u}_3 \\ h_1 dq_1 & h_2 dq_2 & h_3 dq_3 \\ E_1 & E_2 & E_3 \end{vmatrix} = 0 \tag{2.6-9}$$

In order for Eq. (2.6-9) to be satisfied, each component must be zero. In general,

$$h_i dq_i E_{i+1} - h_{i+1} dq_{i+1} E_i = 0 \tag{2.6-10a}$$

Rearranging,

$$\frac{h_i dq_i}{E_i} = \frac{h_{i+1} dq_{i+1}}{E_{i+1}} \tag{2.6-10b}$$

which is the general differential equation for field-direction lines in generalized curvilinear coordinates. Equation (2.6-10b) is actually three differential equations where $i = 1,2,3$ and where $i+1 = 2,3,1$, respectively.

### 2.6.1 Cartesian (rectangular) coordinates

In Cartesian coordinates the metric coefficients are unity and the system of coordinates and metric coefficients become

$$q_1 = x \tag{2.6-11a}$$

$$q_2 = y \tag{2.6-11b}$$

$$q_3 = z \tag{2.6-11c}$$

$$h_1 = 1 \tag{2.6-12a}$$

$$h_2 = 1 \tag{2.6-12b}$$

$$h_3 = 1 \tag{2.6-12c}$$

and Eq. (2.6-10b) becomes

$$\frac{dx}{E_x} = \frac{dy}{E_y} \tag{2.6-13a}$$

$$\frac{dy}{E_y} = \frac{dz}{E_z} \tag{2.6-13b}$$

$$\frac{dz}{E_z} = \frac{dx}{E_x} \tag{2.6-13c}$$

That is, when $E_x$, $E_y$, and $E_z$ are known, the simultaneous solution to the three differential equations (2.6-13a)–(2.6-13c) describes the equations in $(x, y, z)$ space of lines everywhere parallel to the field $\overline{E}$.

### 2.6.2 Cylindrical coordinates

In cylindrical coordinates we may use $(r, \phi, z)$; however, since $r$ is also used for the spherical radius coordinate, which is not the same, it is necessary to distinguish one from the other in problems where both coordinate systems are being used simultaneously. In such cases, we use different symbols in the analysis, such as $r_c$ and $r_s$, for the cylindrical and spherical coordinate radii, respectively. However, since we are treating the coordinates separately, we will just use the symbol "$r$" for each. Thus, for cylindrical coordinates we specify our coordinates and metric coefficients as

$$q_1 = r \tag{2.6-14a}$$

$$q_2 = \phi \tag{2.6-14b}$$

$$q_3 = z \tag{2.6-14c}$$

$$h_1 = 1 \tag{2.6-15a}$$

$$h_2 = r \tag{2.6-15b}$$

$$h_3 = 1 \tag{2.6-15c}$$

and Eq. (2.6-10b) becomes

$$\frac{dr}{E_r} = \frac{rd\phi}{E_\phi} \tag{2.6-16a}$$

$$\frac{rd\phi}{E_\phi} = \frac{dz}{E_z} \tag{2.6-16b}$$

$$\frac{dz}{E_z} = \frac{dr}{E_r} \tag{2.6-16c}$$

Again, when $E_r$, $E_\phi$, and $E_z$ are known, the simultaneous solution to the three differential equations (2.6-16a)–(2.6-16c) describes the equations in $(r, \phi, z)$ space of lines everywhere parallel to the field $\overline{E}$.

### 2.6.3 Spherical coordinates

For spherical coordinates, we specify our coordinates and metric coefficients (as stated earlier) as

$$q_1 = r \tag{2.6-17a}$$

$$q_2 = \theta \tag{2.6-17b}$$

$$q_3 = \phi \tag{2.6-17c}$$

$$h_1 = 1 \tag{2.6-18a}$$

$$h_2 = r \tag{2.6-18b}$$

$$h_3 = r \sin\theta \tag{2.6-18c}$$

and Eq. (2.6-10b) becomes

$$\frac{dr}{E_r} = \frac{r d\theta}{E_\theta} \tag{2.6-19a}$$

$$\frac{r d\theta}{E_\theta} = \frac{r \sin\theta d\phi}{E_\phi} \tag{2.6-19b}$$

$$\frac{r \sin\theta d\phi}{E_\phi} = \frac{dr}{E_r} \tag{2.6-19c}$$

Once more, when $E_r$, $E_\theta$, and $E_\phi$ are known, the simultaneous solution to the three differential equations (2.6-19a) through (2.6-19c) describes the equations in $(r, \theta, \phi)$ space of lines everywhere parallel to the field $\overline{E}$.

### 2.6.4 Example of field direction lines

Let us apply these concepts to the case of a uniformly charged straight line of finite length, where the charge density is $\rho_L$ coulombs per meter lying on the $z$

axis in the range $-a \le z \le a$. The geometry for this configuration is shown in Fig. 2.6-1. The field point expressed in cylindrical coordinates is at *P(r,z)*. The electric field intensity $\overline{E}(r,z)$ is readily found as[15]

$$\overline{E} = \frac{\rho_L}{4\pi\varepsilon r}\left[\hat{u}_r\left(\frac{z+a}{R_1} - \frac{z-a}{R_2}\right) - \hat{u}_z\left(\frac{r}{R_1} - \frac{r}{R_2}\right)\right] \tag{2.6-20}$$

where

$$R_1 = \sqrt{r^2 + (z+a)^2} \tag{2.6-21a}$$

and

$$R_2 = \sqrt{r^2 + (z-a)^2} \tag{2.6-21b}$$

Note that the $\overline{E}$ field is not a function of the azimuthal coordinate $\phi$ nor does it have a $\phi$ component, as expected from the symmetry. Therefore, the field direction lines will lie on surfaces of revolution about the $z$ axis.

From Eqs. (2.6-16c) and (2.6-20), the field direction lines must satisfy the differential equation

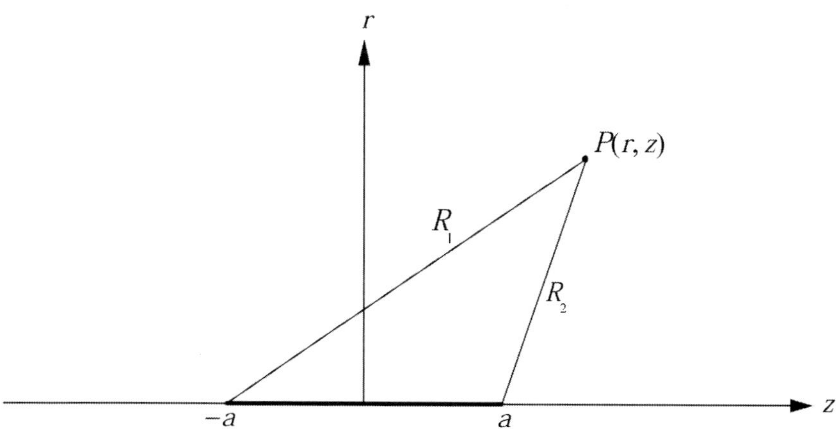


**Figure 2.6-1** Geometry for the field from a uniform, straight line charge $\rho_L$ of finite length 2*a* in cylindrical coordinates.

$$\frac{dr}{dz}=\frac{E_r}{E_z}=\frac{\left(\dfrac{z+a}{R_1}-\dfrac{z-a}{R_2}\right)}{\left(\dfrac{r}{R_2}-\dfrac{r}{R_1}\right)} \tag{2.6-22}$$

where $E_r$ and $E_z$ are the $r$ and $z$ components from Eq. (2.6-20), respectively. Rogers shows[15] that the solution to Eq. (2.6-22) is

$$R_1 - R_2 = C \tag{2.6-23}$$

where $R_1$ and $R_2$ are defined in Fig. 2.6-1. The constant of integration $C$ can be any positive or negative real value including zero. Equation (2.6-23) represents a family of confocal hyperbolas with foci at $z=\pm a$, that is, at the ends of the charged line.

A map of the $\bar{E}$ field direction lines is shown in Fig. 2.6-2 for $C=\sqrt{3}a,\ a,\ 0,\ -a,\ -\sqrt{3}a$. These are hyperbolas with asymptotes at 30, 60, 90, 120, and 150 degrees from the positive $z$ axis, respectively. Rogers[15] (Fig. 2.10) cleverly includes a circle of radius $a$ centered at the center of the charged line from which construction lines may be drawn that define the apexes and asymptotes of each hyperbola. Thus, with the apex and asymptotes known, one can fairly accurately sketch the hyperbolas without the need for detailed calculations.

In 3D space, surfaces on which field direction lines fall can be shown by the simultaneous solution of the differential equations (2.6-16a) and (2.6-16c). We have already discussed the solution of the latter. Let us next examine Eq. (2.6-16a).

$$\frac{r\,d\phi}{E_\phi}=\frac{dr}{E_r} \tag{2.6-24}$$

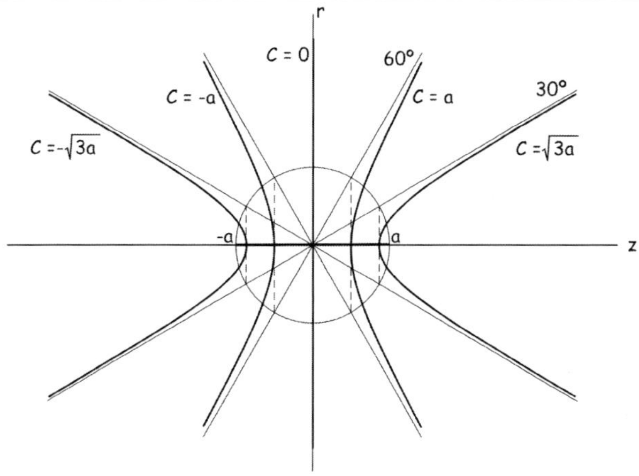


**Figure 2.6-2** Field direction lines of a uniformly charged straight line of length 2*a*.

However, from Eq. (2.6-20), $E_\phi = 0$ since $r \neq 0$, $d\phi$ must equal zero. Integrating we find that $\phi$ equals a constant—independent of the constant of integration given in Eq. (2.6-23). Thus, in 3D space, the equations for the field direction lines must satisfy[15]

$$\begin{aligned} R_1 - R_2 &= C_1 \\ \phi &= C_2 \end{aligned} \tag{2.6-25}$$

The 3D depiction of Eq. (2.6-23) is the hyperboloids of revolution shown in Fig. 2.6-3. For $C = 0$, the surface is in the $z = 0$ plane. For $C > 0$, the surfaces are hyperboloids in the positive $z$ half space, and for $C < 0$, the hyperboloids fall in the negative $z$ half space. Figure 2.6-3 shows the hyperboloids for the four cases, $C = \pm\sqrt{3}\,a$ and $\pm a$. The surfaces that satisfy Eq. (2.6-25) are confocal hyperboloids of revolution with foci at $z = \pm a$ as shown in Fig. 2.6-3. Thus,

*the field direction lines are the lines of intersection between these hyperboloids and any plane containing the z axis.*

## 2.7 Scalar Field Equivalue Surfaces

In the previous section, we developed the process for determining the surfaces of vector field direction lines and illustrated how this works for the case of the electric field $\overline{E}$ from a uniformly charged straight line of finite length. Recall that we did this by forcing an unknown differential tangent vector $\overline{dT}$ to be parallel to $\overline{E}$ by setting $\overline{dT} \times \overline{E}$ to be zero and solving for $\overline{dT}$. In this section, we consider the orthogonal problem, namely that of equivalue surfaces.

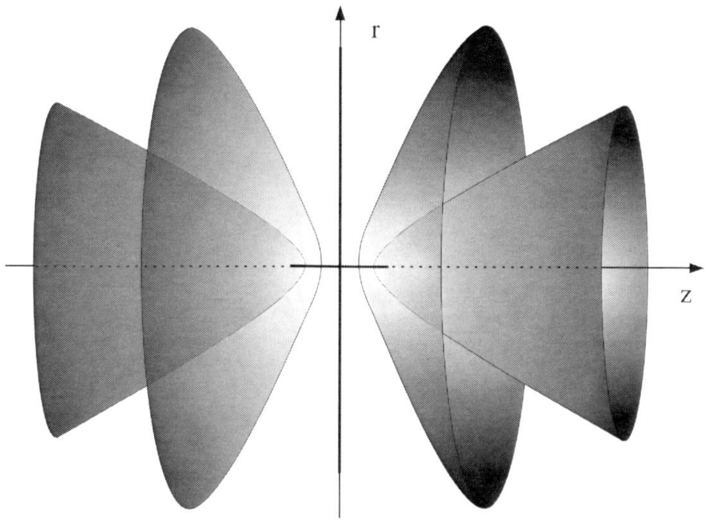


**Figure 2.6-3** Confocal hyperboloids of revolution for the field of a uniformly charged straight line of finite length. The ends of the line charge lie at the foci of the hyperboloids.

If a field source, such as a test mass in a gravitational force field, or a test charge in an electrostatic field (as in the illustration of the previous section), is caused to move in a direction perpendicular to the field direction lines, no work is done. This is because $\overline{F} \cdot \overline{d\ell}$ is zero when $\overline{d\ell}$ is perpendicular to $\overline{F}$. Thus, in any conservative field [see the discussion of Eq. (2.4-22)] there exist surfaces of equal potential energy that are orthogonal to the vector field flow lines. The same can be said of equal electric potential surfaces in the case of electric field intensity $\overline{E}$. Both are called *equipotential surfaces*, one referring to surfaces of equal potential energy and the other referring to surfaces of equal electric potential. We may refer to such surfaces as *equivalue surfaces* in order to generalize our discussion to any conservative vector field.

These equivalue surfaces may be found by first defining a differential path $\overline{dP}$ that is everywhere perpendicular to the vector field, $\overline{E}$, for example. Our task is to determine the differential vector $\overline{dP}$ such that it is everywhere orthogonal to $\overline{E}$. This can be accomplished by noting the following:

*Any two vectors are perpendicular when their dot product is zero.*

That is, by restricting their dot product to be zero, we have the necessary mathematical construct to find $\overline{dP}$ in terms of the given vector field $\overline{E}$. Thus, if the expression

$$\overline{dP} \cdot \overline{E} = 0 \qquad (2.7\text{-}1)$$

is everywhere satisfied, $\overline{dP}$ is everywhere orthogonal to $\overline{E}$. Rogers[15] solves this for the case of the uniformly charged straight line charge of finite length simply as

$$R_1 + R_2 = C \qquad (2.7\text{-}2)$$

where $R_1$ and $R_2$ are defined in Fig. 2.6-1. This simple solution is the form of confocal ellipsoids with foci at the ends of the line at $z = \pm a$, as shown in Fig. 2.7-1. These ellipses are everywhere orthogonal to the confocal hyperbolas shown in Fig. 2.6-2. As was the case there, the solutions are surfaces of revolution about the $z$ axis and, thus, are confocal ellipsoids, as shown in Fig. 2.7-2, that are everywhere orthogonal to the confocal hyperboloids of Fig. 2.6-3. These ellipsoids are the desired equivalue surfaces.

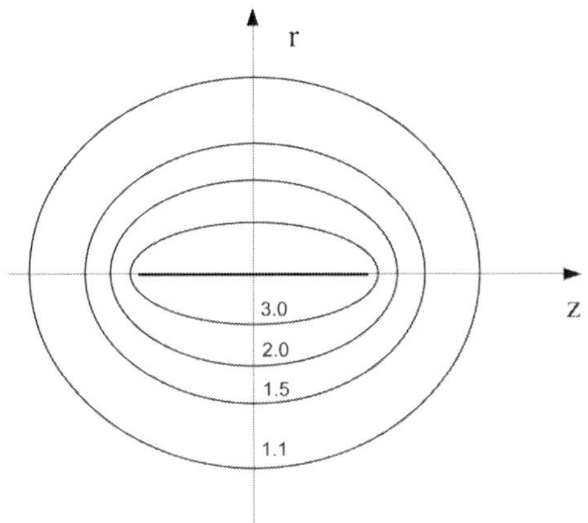


**Figure 2.7-1** Confocal equipotential ellipses for the case of a uniformly charged straight line charge of finite length, where the ends of the line charge lie on the foci of the ellipses.

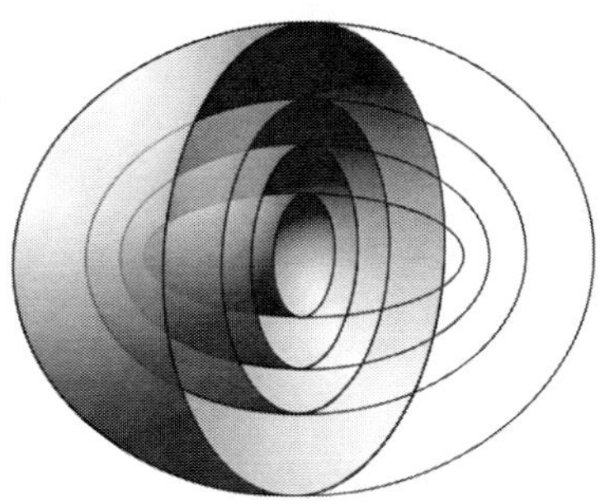

**Figure 2.7-2** Equipotential ellipsoidal surfaces of revolution for the case of Fig. 2.6-1.

## References

1. Simon Ramo, John R. Whinnery, and Theodore Van Duzer, *Fields and Waves in Communication Electronics*, 3rd ed., Wiley, New York (1994).
2. C. D. Cantrell, *Modern Mathematical Methods for Physicists and Engineers*, Cambridge University Press (2000).
3. George B. Thomas, Jr. and Ross L. Finney, *Calculus and Analytic Geometry*, 8th ed., Addison-Wesley, Reading, MA (1992).
4. F. B. Hildebrand, *Advanced Calculus for Engineers*, 5th printing, Prentice-Hall, Englewood Cliffs, NJ (1955).
5. C. R. Wylie, Jr., *Advanced Engineering Mathematics*, McGraw-Hill, New York (1951).
6. James Stewart, *Calculus*, 2nd ed., Brooks/Cole, Monterey, CA (1991).
7. Peter V. O'Neil, *Advanced Engineering Mathematics*, Wadsworth Publishing Co., Belmont, CA, 3rd ed. (1983).
8. R. Courant and D. Hilbert, *Methods of Mathematical Physics*, Wiley Interscience, Hoboken, NJ, 2nd printing (1953).
9. I. S. Sokolnikoff, *Tensor Analysis: Theory and Applications*, Wiley, New York, 3rd printing (1958).
10. C. D. Cantrell, op. cit.
11. Philip M. Morse and Herman Feshbach, *Methods of Theoretical Physics, Part I*, McGraw-Hill, New York (1953).
12. Govind P. Agrawal, *Nonlinear Fiber Optics*, 3rd ed., Academic Press, London (2001).
13. Granino A. Korn and Teresa M. Korn, *Mathematical Handbook for Scientists and Engineers*, McGraw-Hill, New York (1968).
14. A. I. Lur'e, *Three Dimensional Problems of the Theory of Elasticity*, Wiley Interscience, Hoboken, NJ (1964).
15. Walter E. Rogers, *Introduction to Electric Fields*, McGraw-Hill, New York (1954).

## Chapter 3

# Elementary Tensor Analysis

In keeping with the theme of this book, this chapter is presented for the undergraduate-level student and those who teach undergraduates. The author has found that the average electrical engineering junior can begin his or her excursion into tensors if the concepts are presented at the level given here. (It is assumed that a EE junior will have successfully completed 16 to 18 units of math from Calculus I through differential equations and linear algebra.) There is no reason why students of this caliber should not be shown the power of tensors, especially in light of the inevitable shift into photonics for the design of ultrahigh-speed devices and transmission systems. Students of civil and mechanical engineering can also utilize these concepts in their investigation of composite materials, as can their instructors. In fact, one could take the position that there is an even greater need for tensors in these disciplines because stress, strain, and elastic modulus are tensors even for linear isotropic materials.

Many introductory electromagnetics texts, especially those published more recently, make some mention of tensors when discussing anisotropic media.[1–3] For example, some authors use the term "dyadic," the more precise expression for the particular tensor that provides the needed parameters for linear anisotropic media.[4] The mention of tensors is also included in discussions of linear *bianisotropic* media, especially composite materials, covered in more advanced treatises.[5] In addition, the scattering of electromagnetic waves from objects is skillfully treated in texts by the use of the *scattering dyadic*.[6]

Others use the less precise term "tensor," and define it with the nine components of a dyadic. Yet others, especially in earlier works, discuss the nine-component expansion of the anisotropic media without mentioning either tensors or dyadics.* The more advanced a text in electromagnetics is, the more probable it is that dyadic tensors are used to formulate the mathematical description of the physics, regardless of the whether the text is a classic[7,8] or written more recently.[9,10]

---

* In all fairness, however, most of these were written in the days when tensor analysis was found in only the more advanced texts and was deemed out of reach of the undergraduate student. This guide is intended to help bridge this gap.

### *The tensor/dyadic issue*

Those new to tensors will find a possible confusion in the use of the terms "tensor" and "dyadic" in the literature. In some references, the terms are used synonymously. In others, the two terms are very distinct. The bottom line is that dyadics are special cases of tensors, as will become clear in this and later chapters. Since there are many examples where tensors are needed that are not dyadics, it is more common in recent literature to use the term "dyadic" for the nine-component tensor. These comments are made so that those new to tensors (or wanting to brush up on tensors) may understand the variations in terminology found in the literature.

This chapter strives to make the upper-division student more comfortable with the power and value of tensors in situations where the medium is not necessarily "linear, isotropic and homogeneous," as assumed in conventional, more simplistic analyses in physics and engineering. Section 3.1 deals with *directional compoundedness*, *rank*, and *order* of tensors and is intended to give an appreciation for the handling of higher-rank tensors with tensor notation. This is followed in Section 3.2 by a discussion and determination of the number of components of a tensor. For example, the dyadic (which has a rank of 2) will contain $n_d^2$ components, where $n_d$ refers to the dimensionality of the space. We usually work in 3D or 3-space, so $n_d = 3$. Therefore, we find that the dyadic has nine components, which are expanded in Section 3.3.

The dyadic dot product with a vector and the vector dot product with a dyadic are then carried out in Section 3.4. The first is shown to be consistent with matrix multiplication. The second is not. The dot product and double dot product of two dyadics is performed so that the reader may understand the inner-product nature of the dot-product operation and so that the nonreciprocity of some of the operations becomes evident.

The new paradigm of composite materials can properly be studied and analyzed only with the appropriate use of tensor relationships. Section 3.5 illustrates the fundamental relationship between stress and strain (rank-two tensors) and the modulus of elasticity (a rank-four tensor having 81 components in 3D space). Except for the most simplified cases of linear isotropic materials where the forces are aligned in very specific orientations, tensors are essential.

Further, as the discipline of electrical engineering undergoes a profound new paradigm into optical engineering and photonics (fiber optics, acousto-optics, electro-optics, magneto-optics and optoelectronics), the use of tensors has become prevalent. Moreover, as nonlinear optical effects become more common in the design of photonic systems, tensor analysis is essential.[11–17] This is illustrated in Section 3.6.

## 3.1 Directional Compoundedness, Rank, and Order of Tensors

A quantity can have multiple levels of directionality inherent within itself. The level or compoundedness of the directionality of a quantity can be enumerated. For example, since a scalar, such as temperature, pressure, or potential, has no directionality, it can be said to have a *zero directional compoundedness.** Likewise, a vector quantity such as velocity or acceleration inherently has a single direction associated with it, such as $\hat{u}_i$ or $\overline{E}$, and thus it is said to have a *directional compoundedness* of one. As described in the previous chapter, a dyadic requires a two-level directionality. For example, the two-directional-level permittivity dyadic $\overline{\overline{\varepsilon}}$ is required for anisotropic media and each of its nine components require the two-directional-level unit dyad $\hat{u}_i\hat{u}_j$. Thus, we say that a dyadic has a *directional compoundedness* of two.

> *Another term for directional compoundedness is "rank." Thus, a dyadic is a tensor of rank = 2, a vector is a tensor of rank = 1, and a scalar is a tensor of rank = 0.*

Further, it is important to realize at this juncture that tensors do exist that are neither scalars, vectors, nor dyadics. These are tensors having a directional compoundedness greater than two. For example, as mentioned in the introductory comments to this chapter, students of mechanics of materials where stress and strain are studied will be dealing with the modulus of elasticity, which is a tensor of rank four (even though they may not be told this in their sophomore courses in statics and dynamics). This topic is briefly discussed in Section 3.5. An additional example of the need for tensors of rank greater than two is in the study of how optical signals are amplified. Here one needs to know that nonlinear optical effects play an important role in optical amplification. Therefore, it is important to know how to deal with optical waves that drive dielectric materials into the nonlinear regime. To do so requires analyses that use tensors of a higher rank than two, as we will see in Section 3.6.

---

* "*Directional compoundedness*," a term coined by the author, is intended to give those new to tensors a more intuitive feel for the concept of "*rank*."

### *The rank/order issue*

In earlier works dealing with rank-two tensors—that is, dyadics—it was not uncommon to see the term *order* used interchangeably with rank.[18] However, modern usage of the term enumerates *order* as one less than the rank, especially in photonics, optoelectronics, and nonlinear optics.*

$$\textit{Tensor order} = \textit{tensor rank minus one.} \tag{3.1-1}$$

Order is not normally used in referring to a vector; however, if it were used, the vector would be said to have an order of zero. Such a designation would have no useful meaning. (Of course, order would not apply to a scalar.) Thus, the lowest-rank tensor to which order is normally ascribed is a dyadic. A dyadic, therefore, has an order of one. The table at the end of this chapter summarizes rank and order of various tensors.

## 3.2 Tensor Components

In general, the number of components that it takes to describe a tensorial quantity is related to the dimensionality of the problem and the rank of the quantity in accordance with the expression

---

* This use of the term "order" as being one less than the rank is explicit in References [15] page 17, [14] page 16, and [13] page 190, for example. However, it is implicit in most other modern treatments of optics and photonics as though it were self-understood. For example, in Reference [11] pages 25 ff., Reference [12] pages 19 ff., Reference [14] pages 13 ff., Reference [17] pages 5-7, 341-2, and many others, the order denoted in the superscript (in parentheses) is always one less than the rank denoted in tensor notation by the number of indices in the subscript. Furthermore,

***enumerating order as being one less than the rank affords a more natural description of the physics as well as the mathematics.***

For instance, the so-called **third**-order susceptibility, which is a tensor of rank 4, as shown in Section 3.6, in fact, operates on the **cube** of the vector electric field $\bar{E}^3$ (which is a rank-**three** tensor) creating a **third**-harmonic signal coming from $\cos^3 \omega t$ through a **triple** dot-product operation {See References [11] Eq. (2.66), p. 40; [12] Eq. (1.4.5), p. 31; [13] Eq. (7.12), p. 190 and Eq. (7.36), p. 200; [14] Section 2.1.3, Eq. (2.19), pp. 16 ff.; [15] Eq. (1.3.1), p. 18, Section 2.3.1 pp. 39 ff.; [16] Eq. (2.39), p. 39; [17] Eqs. 1.2.8 and 1.2.11, pp.6 and 7}.

Since this guide is directed toward photonics and materials science that use these higher-rank tensors, this author opts for the generally acceptable use of the term *order* as described by Eq. (3.1-1).

$$n_c = n_d^{\,n_R} \tag{3.2-1}$$

where $n_c$ is the number of components, $n_d$ is the dimensionality of the space, and $n_R$ is the rank number. Thus, in 3D space, scalars have

$$n_c = 3^0 = 1 \tag{3.2-2}$$

one component. Vectors have

$$n_c = 3^1 = 3 \tag{3.2-3}$$

three components. Dyadics have

$$n_c = 3^2 = 9 \tag{3.2-4}$$

nine components, triadics have

$$n_c = 3^3 = 27 \tag{3.2-5}$$

27 components, and quadadics, that is, tensors of rank four, have

$$n_c = 3^4 = 81 \tag{3.2-6}$$

81 components, etc.

Therefore, the elastic modulus (Section 3.5) and the third-order susceptibility (Section 3.6), then, have in general 81 components, since they are quadadics.

## 3.3 Dyadics and the Unit Dyad

In many physical problems a constitutive relation is used to relate an intensity field vector with a flux density vector field by a material-dependent parameter. An example of this is the relation between the electric flux density vector $\bar{D}$ and the electric field intensity $\bar{E}$ as

$$\bar{D} = \varepsilon\bar{E} \tag{3.3-1}$$

where $\varepsilon$ is the material parameter. In this case, $\varepsilon$ is referred to as the electric permittivity. In this formulation, it is assumed that the permittivity $\varepsilon$ is an invariant scalar independent of the direction of the applied field. That is, the above expression is for the case of an *isotropic* medium. In addition, Eq. (3.3-1) also assumes that *linear* conditions are at play. That is, the material properties do not change with the amplitude of the applied field and are therefore linear as well as isotropic.

However, if the medium is *anisotropic* (nonisotropic), that is, its dielectric properties differ with respect to direction, the constitutive relation becomes

$$\overline{D} = \overline{\overline{\varepsilon}} \cdot \overline{E} \tag{3.3-2}$$

where $\overline{\overline{\varepsilon}}$ is the anisotropic permittivity and is a dyadic, which, in general, has nine components $\varepsilon_{ij}$ where $i = 1, 2, 3$ and $j = 1, 2, 3$. Again, Eq. (3.3-2) implies linearity. The details of the operation represented by the right-hand side of Eq. (3.3-2) lie in the mathematical description of a dyadic dot product with a vector. This and other dyadic dot products are given in Section 3.4. The case in which material properties vary with the applied field amplitude is the subject of nonlinear analyses and is outlined in Section 3.6.

The dyadic $\overline{\overline{\varepsilon}}$ in Eq. (3.3-2) can be written as

$$\begin{aligned} \overline{\overline{\varepsilon}} &= \hat{u}_1\hat{u}_1\varepsilon_{11} + \hat{u}_1\hat{u}_2\varepsilon_{12} + \hat{u}_1\hat{u}_3\varepsilon_{13} \\ &+ \hat{u}_2\hat{u}_1\varepsilon_{21} + \hat{u}_2\hat{u}_2\varepsilon_{22} + \hat{u}_2\hat{u}_3\varepsilon_{23} \\ &+ \hat{u}_3\hat{u}_1\varepsilon_{31} + \hat{u}_3\hat{u}_2\varepsilon_{32} + \hat{u}_3\hat{u}_3\varepsilon_{33} \\ &= \sum_{i=1}^{3}\sum_{j=1}^{3} \hat{u}_i\hat{u}_j\varepsilon_{ij} \end{aligned} \tag{3.3-3}$$

The nine components $\varepsilon_{ij}$ are scalars representing the proportionality constant between the applied field component $E_j$ in the $\hat{u}_j$ direction and the resultant displacement (flux) component $D_i$ in the $\hat{u}_i$ direction. The *unit dyads* $\hat{u}_i\hat{u}_j$ depict the nine combinations of applied and resultant field direction components.

In writing the bidirectional dyad $\hat{u}_i\hat{u}_j$, there are no operations (such as dot or cross product operations) implied between the unit vectors $\hat{u}_i$ and $\hat{u}_j$. One should think of the unit dyad as a single entity and might prefer to write it as $\hat{\hat{u}}_{ij}$ to emphasize that situation. Thus,

$$\hat{u}_i\hat{u}_j \equiv \hat{\hat{u}}_{ij} \tag{3.3-4}$$

The explicit representation of the nine dyads $\hat{u}_{i\,j}\,\varepsilon_{i\,j}$ represents the components of the dyadic $\bar{\bar{\varepsilon}}$ by the scalar magnitude $\varepsilon_{i\,j}$ in the bidirection $\hat{u}_{i\,j}$.

However, as discussed in Section 1.1.7(b), in tensor analysis it is customary to streamline the notation by dropping the unit dyad and summation signs. Thus, the dyadic [Eq. (3.3-3)] of the form $\sum\sum\hat{u}_{i\,j}\varepsilon_{i\,j}$ is written in *tensor notation* just as

$$\varepsilon_{ij}$$

and the tensor notation for Eq. (3.3-2) is written simply as

$$D_i = \varepsilon_{ij}\,E_j \tag{3.3-5}$$

where the double summation and dot product are implied. Note $\varepsilon_{i\,j}$, as scalar components of the dyadic $\bar{\bar{\varepsilon}}$, is written in this text differently from $\varepsilon_{ij}$ to distinguish the scalar component $\varepsilon_{i\,j}$ from the dyadic $\varepsilon_{ij}$ itself by the spacing between the subscripts. In handwritten communication, however, this distinction would not be obvious. In addition, vectors (such as $\bar{D}$) written in tensor notation are expressed simply as $D_i$ (with only one subscript denoting a rank-one tensor) and are indistinguishable from the scalar components, $D_i$. Thus, the distinction between the vector component in explicit standard notation and the whole vector in tensor notation—both being $D_i$—is not apparent even in electronic communication depicting vectors.

> *Therefore, when encountering variables that are presented with subscripted indices, in order to distinguish between their scalar components and actual tensors themselves, it is necessary to specify at the outset whether an analysis is being performed in tensor notation or in explicit standard notation.*

This caution assumes that such an analysis does not contain both the components of the tensor and the tensor written in tensor notation (with its multi-subscripted indices). Except when the user is being exposed to tensor analysis for the first time, as is assumed in this guide, it is usually not necessary to be dealing with both. Tensor notation is sufficiently powerful and accurate that the expansion into components is implied (and becomes obvious). Thus, the experienced analyst performs tensor analysis operations with much greater ease using tensor notation with no loss in generality. Nevertheless, in this introduction to tensors, we will show the component details of the common tensor operations as a pedagogical tool to help the new tensor analyst garner an appreciation for the use and power of tensors.

The detailed expansion implied by the dot product operation of Eq. (3.3-2), or equivalently, the inner-product operation implied by the right-hand side of Eq. (3.3-5), is made in the next section, where we will find that in performing inner-product operations as in Eq. (3.3-5), the inner subscript $j$ disappears and we are left only with a single subscript $i$. Thus, the result is the quantity $D_i$, which has a *single* directional compoundedness implied by the single subscript, and therefore a rank of one, yielding $\overline{D}$ as a vector.

## 3.4 Dyadic Dot Products

In Section 2.4.3(b) the dot-product operation between two vectors was expanded in explicit standard notation [Eq. (2.4.13)] illustrating the "inner-product" nature of the dot product. This operation was also carried out in tensor notation, which (in conjunction with the Kronecker delta) yielded the same *scalar* result with greater elegance.

In this section we apply the dot product to four combinations involving dyadics. In Section 3.4.1, the two combinations of dot products between a vector and a dyadic are presented—first the dyadic dot product with a vector and then the vector dot product with a dyadic. Both have a *vector* resultant. The details of the dyadic dot product with a vector are given in explicit standard notation in order to illustrate that the inner-product nature of the dot product also applies when dyadics or tensors of general rank are involved. Again, these operations are done with the elegance of tensor notation. The results are then applied to Eq. (3.3-2), which is the constitutive relation between the electric field intensity $\overline{E}$ and the resultant electric flux density $\overline{D}$ in an anisotropic medium described by the electric permittivity dyadic $\overline{\overline{\varepsilon}}$.

The remaining two dot-product operations are the dot and double-dot product of two dyadics. These are carried out in Sections 3.4.2 and 3.4.3, respectively. We will find that the former yields another *dyadic* and the latter yields a *scalar*.

### 3.4.1 Vector-dyadic dot products

In Part (a) we expand on the dyadic dot product with a vector for the case of electric fields in an anisotropic dielectric and illustrate the tensorial properties involved. In part (b) we generalize this case and compare it with the case of a vector dot product with a dyadic. This part concludes with a discussion of the conditions where the two are and are not equal.

### 3.4.1(a) Application of the dyadic-vector dot product for anisotropic dielectrics

In performing the dyadic dot product with the vector, let us take the case of Eq. (3.3-2) as an example. Expanding the right-hand-side of Eq. (3.3-2), we incorporate Eq. (3.3-3) into Eq. (3.3-2) to yield

$$\bar{D} = \bar{\bar{\varepsilon}} \cdot \bar{E} = \begin{bmatrix} \hat{u}_1\hat{u}_1\varepsilon_{11} & \hat{u}_1\hat{u}_2\varepsilon_{12} & \hat{u}_1\hat{u}_3\varepsilon_{13} \\ \hat{u}_2\hat{u}_1\varepsilon_{21} & \hat{u}_2\hat{u}_2\varepsilon_{22} & \hat{u}_2\hat{u}_3\varepsilon_{23} \\ \hat{u}_3\hat{u}_1\varepsilon_{31} & \hat{u}_3\hat{u}_2\varepsilon_{32} & \hat{u}_3\hat{u}_3\varepsilon_{33} \end{bmatrix} \cdot \begin{bmatrix} \hat{u}_1 E_1 \\ \hat{u}_2 E_2 \\ \hat{u}_3 E_3 \end{bmatrix} \tag{3.4-1}$$

There are 27 dot-product terms resulting from each of the nine components of the dyadic operating on the three components of the vector. However, applying the Kronecker delta [Eq. (2.4-12)] eliminates 18 of these terms, yielding the nine terms shown below. Therefore, this dot product operation can be correctly formulated by following the same rules as the matrix multiplication of a 3 × 3 matrix with a 3 × 1 column matrix with careful attention given to the order of the unit vector dot products. Thus, we have

$$\begin{aligned} \bar{D} = \; & \hat{u}_1\hat{u}_1 \cdot \hat{u}_1\varepsilon_{11}E_1 + \hat{u}_1\hat{u}_2 \cdot \hat{u}_2\varepsilon_{12}E_2 + \hat{u}_1\hat{u}_3 \cdot \hat{u}_3\varepsilon_{13}E_3 \\ + \; & \hat{u}_2\hat{u}_1 \cdot \hat{u}_1\varepsilon_{21}E_1 + \hat{u}_2\hat{u}_2 \cdot \hat{u}_2\varepsilon_{22}E_2 + \hat{u}_2\hat{u}_3 \cdot \hat{u}_3\varepsilon_{23}E_3 \\ + \; & \hat{u}_3\hat{u}_1 \cdot \hat{u}_1\varepsilon_{31}E_1 + \hat{u}_3\hat{u}_2 \cdot \hat{u}_2\varepsilon_{32}E_2 + \hat{u}_3\hat{u}_3 \cdot \hat{u}_3\varepsilon_{33}E_3 \end{aligned} \tag{3.4-2}$$

Notice that the interior dot products of unit vectors are all unity (with the application of the Kronecker delta). Thus, the result of the dot product operation leaves single directionality in each of the nine terms.

However, before performing this step, it is instructive to point out that the form of Eq. (3.4-2) explicitly illustrates the key tensor algebra issues of the right-hand side of Eq. (3.4-2) as follows:

1. *First, each term of Eq. (3.4-2) displays the two-level directionality of the dyadic, the single directionality of the vector, and the dot operation between these quantities.*

2. *The nature of the inner product is explicitly displayed, thereby illustrating why the mathematical term "inner product" is used to depict the vector dot-product operation.*

3. *The application of the inner product eliminates two of the unit vectors in each term, thereby reducing the sum of the ranks of the two quantities involved by two.*

These observations are also valid for tensors of any rank and for multiple inner products.

Combining the terms in Eq. (3.4-2) for each unit vector $\hat{u}_1, \hat{u}_2, \hat{u}_3$ yields

$$\begin{aligned} \overline{D} \;=\; & \hat{u}_1 (\varepsilon_{11} E_1 + \varepsilon_{12} E_2 + \varepsilon_{13} E_3) \\ + & \hat{u}_2 (\varepsilon_{21} E_1 + \varepsilon_{22} E_2 + \varepsilon_{23} E_3) \\ + & \hat{u}_3 (\varepsilon_{31} E_1 + \varepsilon_{32} E_2 + \varepsilon_{33} E_3) \end{aligned} \tag{3.4-3}$$

Equation (3.4-3) then is the result of our explicit expansion of the dyadic dot product with a vector, which yields another vector $\overline{D}$, and, of course, is generally not oriented in the same direction as the force field $\overline{E}$. Equation (3.4-3) is the form generally given in electromagnetic texts for the constitutive relation between the electric flux density and the electric field intensity for anisotropic dielectrics.

The entire process described in the development of Eqs. (3.4-1) through (3.4-3), when written in tensor notation, is simply

$$D_i = \varepsilon_{ij} E_j \tag{3.4-4}$$

where $i = 1, 2, 3$; $j = 1, 2, 3$. Notice that the inner product index $j$ is eliminated in the dot product operation and the only remaining index is $i$, leaving the resultant quantity $D$ as a rank-one tensor—or a vector—as stated before.

### 3.4.1(b) Comparison of the dyadic-vector dot product with the vector-dyadic dot product

In general the inner product between a dyadic $\overline{\overline{A}}$ and a vector $\overline{B}$ in explicit standard notation takes the form of Eq. (3.4-3)

$$\begin{aligned}\bar{\bar{A}}\cdot\bar{B} = &\ \hat{u}_1(A_{11}B_1 + A_{12}B_2 + A_{13}B_3)\\ &+ \hat{u}_2(A_{21}B_1 + A_{22}B_2 + A_{23}B_3)\\ &+ \hat{u}_3(A_{31}B_1 + A_{32}B_2 + A_{33}B_3)\end{aligned} \tag{3.4-5}$$

which is, of course a vector. In tensor notation, Eq. (3.4-5) is

$$\bar{\bar{A}}\cdot\bar{B} = A_{ij}B_k\delta_{jk} = A_{ij}B_j = C_i \tag{3.4-6}$$

where its vector form is denoted by the single subscript of the resultant $C_i$.

However, in reversing the operation—that is, taking the inner product between the vector $\bar{B}$ and the dyadic $\bar{\bar{A}}$—in explicit standard notation, the setup is different:

$$\bar{B}\cdot\bar{\bar{A}} = \begin{bmatrix}\hat{u}_1B_1\\ \hat{u}_2B_2\\ \hat{u}_3B_3\end{bmatrix}\cdot\begin{bmatrix}\hat{u}_1\hat{u}_1A_{11} & \hat{u}_1\hat{u}_2A_{12} & \hat{u}_1\hat{u}_3A_{13}\\ \hat{u}_2\hat{u}_1A_{21} & \hat{u}_2\hat{u}_2A_{22} & \hat{u}_2\hat{u}_3A_{23}\\ \hat{u}_3\hat{u}_1A_{31} & \hat{u}_3\hat{u}_2A_{32} & \hat{u}_3\hat{u}_3A_{33}\end{bmatrix} \tag{3.4-7}$$

Here, the matrix multiplication analogy cited after Eq. (3.4-1) is invalid. Examination of Eq. (3.4-7) shows that of the 27 dot product combinations, only the row-by-row dot products survive. These are then

$$\begin{aligned}\bar{B}\cdot\bar{\bar{A}} = &\ \hat{u}_1\cdot\hat{u}_1\hat{u}_1B_1A_{11} + \hat{u}_1\cdot\hat{u}_1\hat{u}_2B_1A_{12} + \hat{u}_1\cdot\hat{u}_1\hat{u}_3B_1A_{13}\\ &+ \hat{u}_2\cdot\hat{u}_2\hat{u}_1B_2A_{21} + \hat{u}_2\cdot\hat{u}_2\hat{u}_2B_2A_{22} + \hat{u}_2\cdot\hat{u}_2\hat{u}_3B_2A_{23}\\ &+ \hat{u}_3\cdot\hat{u}_3\hat{u}_1B_3A_{31} + \hat{u}_3\cdot\hat{u}_3\hat{u}_2B_3A_{32} + \hat{u}_3\cdot\hat{u}_3\hat{u}_3B_3A_{33}\end{aligned} \tag{3.4-8}$$

Notice that after applying the Kronecker delta to Eq. (3.4-8), thus eliminating the inner products and reducing the rank of each term from three to one as before, the resultant vector components are collected by columns in Eq. (3.4-8). That is, the first column is the $\hat{u}_1$ component, etc. Collecting components, we have

$$\begin{aligned}\bar{B}\cdot\bar{\bar{A}} = &\ \hat{u}_1(B_1A_{11} + B_2A_{21} + B_3A_{31})\\ &+ \hat{u}_2(B_1A_{12} + B_2A_{22} + B_3A_{32})\\ &+ \hat{u}_3(B_1A_{13} + B_2A_{23} + B_3A_{33})\end{aligned} \tag{3.4-9}$$

which is definitely not the same as Eq. (3.4-5), unless the components of $\bar{\bar{A}}$ are symmetrical.

In tensor notation $\bar{B} \cdot \bar{\bar{A}}$ is more simply determined by

$$\bar{B} \cdot \bar{\bar{A}} = B_i A_{jk} \delta_{ij} = B_i A_{ik} = D_k \tag{3.4-10}$$

which is a different vector $D_k = \bar{D}$ from the result $C_i = \bar{C}$ of Eq. (3.4-6). Therefore,

$$\begin{aligned} \bar{\bar{A}} \cdot \bar{B} &\neq \bar{B} \cdot \bar{\bar{A}} \text{ if } A_{ij} \neq A_{ji} \\ \bar{\bar{A}} \cdot \bar{B} &= \bar{B} \cdot \bar{\bar{A}} \text{ if } A_{ij} = A_{ji} \end{aligned} \tag{3.4-11}$$

That is,

*If the dyadic components are symmetrical, the dyadic-vector dot product is commutative; otherwise, it is not.*

### 3.4.2 Dyadic-dyadic dot and double-dot products

The inner product between two dyadics $\bar{\bar{A}}$ and $\bar{\bar{B}}$ then becomes

$$\bar{\bar{A}} \cdot \bar{\bar{B}} = A_{ij} B_{k\ell} \delta_{jk} = A_{ij} B_{j\ell} = E_{i\ell} \tag{3.4-12}$$

which is a new dyadic $E_{i\ell} = \bar{\bar{E}}$. However, the reverse operation $\bar{\bar{B}} \cdot \bar{\bar{A}}$ is

$$\bar{\bar{B}} \cdot \bar{\bar{A}} = B_{k\ell} A_{ij} \delta_{\ell i} = B_{k\ell} A_{\ell j} = F_{kj} \tag{3.4-13}$$

which is a dyadic $F_{kj} = \bar{\bar{F}}$ but not the same as in Eq. (3.4-12).

The double inner product between $\bar{\bar{A}}$ and $\bar{\bar{B}}$ becomes

$$\bar{\bar{A}} : \bar{\bar{B}} = A_{ij} B_{k\ell} \delta_{jk} \delta_{i\ell} = A_{ij} B_{ji} = G \tag{3.4-14}$$

where all unit vectors drop out ending with a scalar $G$. The details of this in explicit standard notation follow.

$$\overline{\overline{A}} : \overline{\overline{B}} = \sum_{i=1}^{3}\sum_{j=1}^{3} A_{ij}B_{ji} =$$

$$\begin{aligned} &= A_{11}B_{11} + A_{12}B_{21} + A_{13}B_{31} \\ &+ A_{21}B_{12} + A_{22}B_{22} + A_{23}B_{32} \\ &+ A_{31}B_{13} + A_{32}B_{23} + A_{33}B_{33} = G \end{aligned} \qquad (3.4\text{-}15)$$

which is a scalar. It is apparent from Eq. (3.4-15) that the double-dot product between two dyadics is commutative:

$$\overline{\overline{A}} : \overline{\overline{B}} = \overline{\overline{B}} : \overline{\overline{A}} \qquad (3.4\text{-}16)$$

## 3.5 The Rank-Four Elastic Modulus Tensor

In order to illustrate the importance of tensors of varying rank in science and engineering, especially where tensors having ranks higher than dyadics are needed, we will briefly discuss a case in structural properties of materials. In mechanics, *stress*—which represents force per unit area throughout the material—can have one set of values in tension and another set of values in shear. This is true even for isotropic materials. Therefore, stress $s$ must in general be represented by a dyadic. The resulting deformation—strain—that ensues is also dependent on whether the force is in tension or shear. Thus, strain $\delta$ is also a dyadic. The necessity for the dual directional compoundedness is even more apparent when one considers stress and strain being applied to anisotropic materials.

As with vector force fields being applied to materials and their corresponding flux fields, a constitutive relation exits between stress and strain. The components of stress are determined by multiplicative operations of the elastic modulus $\mathcal{E}$ on the strain. Stress and strain being tensors of rank two, the modulus is a tensor of rank four, which we will call a quadadic.[19] Thus, for a complex crystalline material and arbitrarily directed applied forces, the elastic modulus must be described in terms of a tensor of rank four (a quantity that has a directional compoundedness of four). The components of this rank-four tensor are separated into 81 components each having a magnitude and a unique rank-four unitary tensor. Since unit tensors of rank two are called *dyads*, and unit tensors of rank three are *triads*, we refer to the unit tensor of rank four as a *quadad*. Thus, a quadad consists of four unit vectors $\hat{u}_i\hat{u}_j\hat{u}_k\hat{u}_\ell$ run together (not separated by dot or cross operations). To punctuate the concept that quadads are single quantities with four inherent directions at every point in space, we may alternatively express the quadad as a $u$ with four hats or chevrons:

$$\hat{\hat{\hat{\hat{u}}}}_{ijk\ell}$$

The tensor notation that fully represents this rank-four unitary quantity is, as one might expect from Section 3.3, more simply written as

$$u_{ijk\ell}$$

without the cumbersomeness of the chevrons and the implied quadruple summations. The latter representation is customarily used in tensor analysis. In this example, one can begin to see the great utility and simplicity in using tensor notation, although, in doing so one must be careful to order any inner-product operations properly.

Now that the unit quadads are described, we may then express the classical constitutive relation between stress and strain and elastic modulus as

$$\overline{\overline{s}} = \overline{\overline{\overline{\overline{\varepsilon}}}} : \overline{\overline{\delta}} \tag{3.5-1}$$

where $\overline{\overline{s}}$ is the stress dyadic, $\overline{\overline{\delta}}$ is the strain dyadic, $\overline{\overline{\overline{\overline{\varepsilon}}}}$ is the rank-four quadadic tensor modulus of elasticity, and the ":" represents a double dot product (Section 3.4.2) or two "inner product" operations of the elastic modulus upon the strain. Tensor notation for Eq. (3.5-1) is

$$s_{ij} = \varepsilon_{ijk\ell}\delta_{\ell k} \tag{3.5-2}$$

Note that the first dot or inner-product operation eliminates the inside index $\ell$ and the second dot or inner-product operation eliminates the next inside index $k$. The resulting quantity is therefore left with the double indices $ij$, and, thus, the stress dyadic $s_{ij}$ has a dual directional compoundedness or rank of two.

With the tensor notation of Eq. (3.5-2), the double dot product, dyads, and quadads are implied, greatly simplifying the appearance of the operation but maintaining its internal richness.

In the most general case, namely where the material is anisotropic and the stress is applied generally, all 81 components are nonzero. However, even in this most general case, only 21 of the 81 components have different values. For the case of isotropic materials, 60 components are zero and the stress/strain relationship reduces to a six-dimensional vector. This, in turn, reduces to six scalar equations, which are often presented to sophomores without mention of the tensorial origin.

## 3.6 The Use of Tensors in Nonlinear Optics

Another example of the power of tensors is in nonlinear optics. In linear optics (and electromagnetics), the constitutive relation describing the electric flux vector $\bar{D}$ in terms of the applied electric field intensity $\bar{E}$ is given by Eq. (3.3-2), namely $\bar{D} = \bar{\bar{\varepsilon}} \cdot \bar{E}$ It is clear from Eq. (3.3-2) that, if $\bar{E}$ doubles, $\bar{D}$ also doubles (even though two fields are not necessarily parallel). In other words, none of the nine components of $\bar{\bar{\varepsilon}}$ change with $\bar{E}$. This is what is meant in describing the medium as being *linear*.

In contrast the term *nonlinear* is used in optics when the magnitude of the $\bar{E}$ field is high enough to drive the medium into nonlinearity. That is, if the strength of the $\bar{E}$ field is sufficient to change at least one component of $\bar{\bar{\varepsilon}}$, the medium is said to be nonlinear. Nonlinearities in materials take on many forms. These can be grouped into two categories: those that cause permanent changes to the material and those that retain their original properties after the force field—in this case the $\bar{E}$ field—is removed.

The phenomenon of *dielectric breakdown* is an example of the first category. In this case, the $\bar{E}$ field exceeds the *dielectric strength*[20] of a material, thus permanently and deleteriously breaking down the dielectric material and changing its properties in such a way as to render the material ineffective as a dielectric insulator. Another example is the case of an optical fiber becoming irradiated with excessively strong x-rays—the fiber becomes "gray" and its attenuation properties are increased to the point of rendering the fiber inoperable.

The second category of nonlinearities in optical materials occurs whenever the applied $\bar{E}$ field has sufficient strength to change the properties of the material, without the strength to prevent the material from returning to its original nonstressed state. In nonlinear optics, the electric flux density $\bar{D}$ field is given as a series of increasing powers of $\bar{E}$. With each additional term, higher rank permittivity tensors operate on the powers of $\bar{E}$ with increasing inner product operations:

$$\bar{D} = \bar{\bar{\varepsilon}} \cdot \bar{E} + \bar{\bar{\bar{\varepsilon}}} : \bar{E}\bar{E} + \bar{\bar{\bar{\bar{\varepsilon}}}} \vdots \bar{E}\bar{E}\bar{E} + \cdots \quad (3.6\text{-}1)$$

The first term on the right-hand side of Eq. (3.6-1) represents the linear part of the $\overline{D}$ field consistent with Eq. (3.3-2). This term is also referred to as the *first order* contribution to the constitutive relation for two reasons. First, the $\overline{E}$ field is raised to the *first* power. Secondly, there is a *single* inner product. Thirdly, the (rank-two) permittivity dyadic $\overline{\overline{\varepsilon}}$ is of *first* order as defined by Eq. (3.1-1). See the footnote for the usual nonlinear optics representations in terms of susceptibility tensors.*

As in the discussion following Eq. (3.3-3) regarding the bidirectional dyad $\hat{u}_i\hat{u}_j$, we note that there is no dot or cross product operation implied between the two electric field intensities in the second term of the right-hand side of Eq. (3.6-1). Therefore, the quantity $\overline{E}\overline{E}$ is a dyadic or tensor of rank two. Furthermore, the two electric fields in the dyadic $\overline{E}\overline{E}$ are in general not necessarily in the same direction. Likewise, the $\overline{E}\overline{E}\overline{E}$ term of Eq. (3.6-1) is a *triadic* or tensor of rank three.

Beyond the linear, first-order term of the right-hand side of Eq. (3.6-1), the remaining terms represent the optical nonlinearity and are given in terms of increasing powers of $\overline{E}$. The coefficients of these higher powers of $\overline{E}$ are the increasing order permittivities operating on the $\overline{E}^n$s (where $n = 2, 3, \cdots$) through $n$ inner-product operations. For example, in the second term of the right-hand side of Eq. (3.6-1), the triadic permittivity

$$\overline{\overline{\overline{\varepsilon}}} = \varepsilon^{(2)} \tag{3.6-2}$$

operates on the $\overline{E}\overline{E}$ dyadic through a double dot-product operation denoted by ":" as in Eq. (3.5-1) between the *quadadic* modulus and the dyadic strain. This second term is referred to as the *second-order nonlinearity* or the *second-order permittivity*. This term is important for materials that lack *molecular inversion symmetry*, also known as *non-centrosymmetric materials*. Notice that the *second-*

---

* In nonlinear optics, it more convenient to expand the polarization vector $\overline{P}$ rather than the flux-density vector as in Eq. (3.6-1). Thus, $\overline{D} = \varepsilon_0 \overline{E} + \overline{P}_L + \overline{P}_{NL}$, where $\overline{P}_L$ is the linear polarization vector given by $\overline{P}_L / \varepsilon_0 = \overline{\overline{\chi}} \cdot \overline{E}$, where $\overline{\overline{\chi}}$ is the linear anisotropic electric susceptibility dyadic, called the first-order susceptability. $\overline{P}_{NL}$ is the nonlinear polarization given by $\overline{P}_{NL} / \varepsilon_0 = \overline{\overline{\overline{\chi}}} : \overline{E}\overline{E} + \overline{\overline{\overline{\overline{\chi}}}} \vdots \overline{E}\overline{E}\overline{E} + \cdots$, where $\overline{\overline{\overline{\chi}}}$ and $\overline{\overline{\overline{\overline{\chi}}}}$ are the second- and third-order nonlinear susceptibilities which are tensors of rank 3 and 4, respectively. Nevertheless, Eq. (3.6-1) illustrates the methodology for dealing with nonlinear phenomena and the need and power of using dyadics and tensors of higher rank than dyadics.

order nonlinearity is the result of a *second*-order permittivity $\varepsilon^{(2)}$ operating through a *double* dot product on $\overline{E}$ raised to the power of *two*.

In the third term of Eq. (3.6-1), the *quadadic* permittivity $\overline{\overline{\overline{\overline{\varepsilon}}}}$ operates on the $\overline{E}\overline{E}\overline{E}$ triadic through the triple-dot product operation denoted by "$\vdots$". Thus, this third term is referred to as the *third-order nonlinearity*. Notice further that the *third*-order nonlinearity is the result of a *third*-order permittivity, which is a tensor of rank four, shown as

$$\overline{\overline{\overline{\overline{\varepsilon}}}} = \varepsilon^{(3)} \tag{3.6-3}$$

operating through a *triple*-dot product on $\overline{E}$ raised to the *third* power. The third-order nonlinearity plays an important role for all dielectric materials operating in the nonlinear regime whether the material is centrosymmetric or not.

### *Applications: Optical amplification and soliton waves*

Whereas nonlinear operation can have several deleterious effects on optical signals, not all effects are negative. For example, *Raman amplification*, which is a method for directly amplifying optical signals, occurs when the information-carrying signal interacts through weak nonlinear coupling with a strong pumping signal of higher quantum energy (higher frequency). Another example of favorable nonlinear effects is in *soliton wave propagation*. Solitons are special kinds of wave packets that can travel undistorted over long distances as a result of two offsetting effects: A nonlinear effect called *self-phase modulation* (SPM) tends to offset a linear effect called *group velocity dispersion* (GVD) under special conditions that can be designed into an optical communication system by the lightwave systems designer.

Nonlinear effects, desired or not, occur in optical fibers in the natural process of the design of lightwave systems. On the one hand, there is a need to have enough signal strength so that the signal can be detected at the receiving end of the fiber after the inevitable attenuation (small as it is). On the other hand, the medium through which this signal must propagate is restricted in size by the fiber size, typically 8 μm (microns or micrometers) in diameter. These two factors in the design usually lead to signal intensities that cause the material properties to vary. The designer must avoid at all costs the breakdown phenomenon mentioned in the third paragraph of this section. Since there are usually two or so orders of magnitude between the onset of nonlinearities and actual material breakdown, there is a possibility that higher-order terms of Eq. (3.6-1) may need to be considered although each term falls off in strength quite rapidly. Again, if the material is noncentrosymmetric, the fourth-order nonlinearity is the next term to be considered. However, for centrosymmetric materials, such as the silica glass typically used in optical fibers, the next higher-order term of Eq. (3.6-1) would be the fifth-order nonlinearity. That is, for silica glass, only the odd-order terms of Eq. (3.6-1) need to be considered.

From the above discussion, which recognizes that the order of a tensor is one less than its rank by definition [see 3.1-1], we can express Eq. (3.6-1) in *order notation* a little more simply as

$$D = \varepsilon^{(1)} E + \varepsilon^{(2)} E^2 + \varepsilon^{(3)} E^3 + \cdots \tag{3.6-4}$$

or even more simply as

$$D = \sum_{n=1,2,\cdots} \varepsilon^{(n)} E^n \tag{3.6-5}$$

In doing so, however, one must realize that there are inner-product operations implicit between the permittivity tensors and the $E$-field powers, and that the latter are true tensors.

The *tensor notation* for Eq. (3.6-1) is

$$D_i = \varepsilon_{ij} E_j + \varepsilon_{ijk} E_k E_j + \varepsilon_{ijk\ell} E_\ell E_k E_j + \cdots \tag{3.6-6}$$

where, again, the inner-product operations are implied. As usual, the number of subscripts denotes the rank of the respective tensor; however, the order of the subscripts on the $E$-field vectors is carefully chosen to denote the order of the inner-product operations. The disappearance of the inner indices in inner-product operations was illustrated in detail in Eq. (3.4-2) in the development of the linear term of Eq. (3.6-6), namely Eq. (3.3-5), where the $j$ index was eliminated as a result of the single inner-product operation.

The same phenomenon takes place in the second-order nonlinear term in Eq. (3.6-6) with the elimination of the $k$ index, and then, the $j$ index from the double inner product leaving only the $i$ index. In the third-order nonlinear term, the inner indices are again eliminated—first the $\ell$ index, then the $k$ index, and finally the $j$ index, resulting from the triple inner-product operation, leaving only the $i$ index again. The process continues if higher-order nonlinearities are used. In all cases, all inner-product indices are eliminated leaving only the $i$ index, which is consistent with the vector form of the left-hand side of Eq. (3.6-6). Equations (3.6-1) and (3.6-4) through (3.6-6) are formulations of the same constitutive relation, the first term of each representing the linear component and the remaining terms representing the nonlinear components.

*Summary of the use of tensors for nonlinear optics:*

Once the nonlinear formulation is set up, the remaining task is to establish the components of the tensorial coefficients $\varepsilon^{(n)}$. This is usually accomplished empirically by carefully constructed experimental procedures. The resulting values are dependent upon the material. For example, for the case of silica ($SiO_2$) fiber, the second-order permittivity vanishes, that is, $\overline{\overline{\overline{\varepsilon}}} = \varepsilon^{(2)} = \varepsilon_{ijk} = 0$, leaving the third-order permittivity as the lowest-order nonlinearity.

## 3.7 Term-by-Term Rank Consistency and the Rules for Determining Rank after Performing Inner-Product Operations with Tensors

Every scientist and engineer knows that each term of an equation must have consistent units. You cannot add volts and amps. The same rule applies to rank. Each term of an equation must have the same rank. In order to determine the rank of a quantity involving multiplications of tensors, the rule is to add the ranks of each tensor being multiplied. If the tensors are undergoing inner-product operations, the rule is to subtract two in rank* for each inner-product operation.

*Example 1: The electric field constitutive relation*

To find the resultant rank after performing the single inner product operation on the right side of Eq. (3.3-5), one must first add the ranks of the tensors, namely two and one for the dyadic and vector, respectively, and subtract two for the single inner product. Thus, we have $2+1-2=1$ and our resultant parameter $\overline{D}$ is a tensor of rank one, which is a vector.

*Example 2: Materials mechanics constitutive relation*

Likewise, the resultant from our double-dot product operation on the right side of Eq. (3.5-1) is 4 + 2: the quadadic (tensor of rank four) modulus of elasticity plus

---

* Another rule is to subtract one in rank for each cross-product operation in the term; however, since this introductory chapter on tensors omits such operations, examples of this rule are given in Chapters 4 and 5.

dyadic strain (tensor of rank two) minus two for each of the two inner products. Thus, $4+2-2-2=2$ and our resultant is therefore a tensor of rank two, which, in this example, is the stress dyadic.

*Example 3: Nonlinear optics*

Further, as implied in the discussion of the elimination of inner indices of Eq. (3.6-6), the second term involves a triadic and a dyadic, having combined ranks of $3+2$, with two inner products reducing the rank by $-2-2$, or $3+2-2-2=1$. The third term involves a quadadic and a triadic less three inner products yielding $4+3-2-2-2=1$. Thus, each term of Eq. (3.6-6) reduces to tensors of rank one, i.e. each term reduces to a vector—compatible with the vector field $\bar{D}$ on the left-hand side.

## 3.8 Summary of Tensors

The objective of this chapter has been to introduce some of the basic elements of tensor analysis to those who are new to tensors and their uses in describing phenomena that otherwise cannot be properly described by using ordinary vector formulations. When first introduced to vectors, students are told that, whereas

*scalars are quantities that have magnitude but no direction,*

a quantity that has magnitude and direction is a vector. We now find that quantities that have magnitude and direction may, in fact, not be vectors but may be tensors having multiple directional compoundedness. Thus, the definition of a vector must have the word "single" inserted, to wit:

*A quantity that has magnitude and a* ***single*** *direction is a vector.*

The same kind of philosophical leap in contemplating a vector field after understanding a scalar field is needed in order to contemplate a dyadic field after understanding a vector field—namely the concept of two inherent directions at every point in the field from the concept of one inherent direction. Those new to tensors are cautioned not to stop there, because, by so doing one might be lead to the conclusion that tensors are dyadics. Whereas dyadics are tensors (of rank two, specifically) *tensors are not in general dyadics*. The terms are not synonymous. Dyadics are quantities that have two directions associated with each point in the field. Triadics have three directions, quadadics have four, etc. Thus,

*A tensor is a quantity that has multiple directionality at each point in space and at each moment in time. The "rank" of a tensor enumerates that multiplicity.*

Therefore,

*a scalar is a tensor of rank zero,*

*a vector is a tensor of rank one,*

*a dyadic is a tensor of rank two,*

*a triadic is a tensor of rank three,*

*a quadadic is a tensor of rank four,*

*etc.*

Furthermore, tensors can be operated upon like the familiar vector arithmetic operations, such as dot, cross, and direct products.* The resultant quantity after performing such operations between them can be summarized as

*a direct product of two tensors yields another tensor of rank equaling the sum of the ranks of the two tensors,*

*the cross product of two tensors yields a tensor of rank one less than the sum of the ranks of the two tensors,*

*the dot product of two tensors yields a tensor of rank two less than the sum of the ranks of the two tensors, and*

*the multiple dot product of two tensors yields a tensor of rank equaling the sum of the ranks of the two tensors less two for each multiple dot product.*

The following table summarizes tensor terminology and some of the topics covered in this chapter.

---

* See footnote on page 2-7.

**Table 3-1** Tensor Nomenclature and Applications

| Field name | Alternative Name | Unit name | Tensor Rank | Tensor Order | Applications (selected examples) |
|---|---|---|---|---|---|
| scalar | | | 0 | n/a | temperature; pressure; voltage; current; flux |
| vector | | unit vector | 1 | (not used) | velocity; acceleration; electric, magnetic field intensity; electric, magnetic flux density |
| dyadic | "tensor" (poor usage) | dyad | 2 | 1st | stress; strain; linear anisotropic, bianisotropic permittivity; permeability; electric, magnetic susceptibilities |
| triadic | | triad | 3 | 2nd | lowest-order nonlinear susceptibility ( $\chi_e^{(2)}$ ) for non-centrosymmetric materials |
| quadadic | tetradic | quadad | 4 | 3rd | Elastic modulus; lowest-order nonlinear susceptibility ( $\chi_e^{(3)}$ ) for centrosymmetric materials |
| general tensor | tensor | unit tensor | $n_R$ | $n_R - 1$ | all of the above and more |

## References

1. Liang Chi Shen and Jin Au Kong, *Applied Electromagnetism*, 3rd ed., PWS Publishing, Boston, MA (1995).
2. Nannapaneni Narayana Rao, *Elements of Engineering Electromagnetics*, 5th ed., Prentice-Hall, Englewood Cliffs, NJ (2000).
3. Matthew N. O. Sadiku, *Elements of Electromagnetics*, 3rd ed., Oxford University Press (2001).
4. Jeffery L. Young, "Finite-difference time-domain methodologies for electromagnetic wave propagation in complex media," in *Frontiers in*

*Electromagnetics*, Douglas H. Werner and Raj Mittra, Eds., IEEE Press, Piscataway, NJ (2000).

5. Mustafa Kuzuoglu and Raj Mittra, "A systematic study of perfectly matched absorbers," in *Frontiers in Electromagnetics*, Douglas H. Werner and Raj Mittra, Eds., IEEE Press, Piscataway, NJ(2000).
6. Carl E. Baum, "Target symmetry and the scattering dyadic," in *Frontiers in Electromagnetics*, Douglas H. Werner and Raj Mittra, Eds., IEEE Press, Piscataway, NJ (2000).
7. Julius Adams Stratton, *Electromagnetic Theory*, McGraw-Hill, New York (1941).
8. Philip M. Morse and Herman Feshbach, *Methods of Theoretical Physics*, McGraw-Hill, New York (1953).
9. R. E. Collin, *Field Theory of Guided Waves*, 2nd ed., IEEE Press, Piscataway, NJ (1991).
10. Frederic Mariotti, Bruno Sauviac, and Sergei A. Tretyakov, "Artificial bianisotropic composites," in *Frontiers in Electromagnetics*, Douglas H. Werner and Raj Mittra, Eds., IEEE Press, Piscataway, NJ (2000).
11. Y. R. Shen, *The Principles of Nonlinear Optics*, Wiley, New York (1984).
12. Robert W. Boyd, *Nonlinear Optics*, Academic Press, London (1992).
13. Clifford R. Pollock, *Fundamentals of Optoelectronics*, Richard D. Irwin, Inc., Burr Ridge, IL (1995).
14. P. N. Butcher & D. Cotter, *The Elements of Nonlinear Optics*, Cambridge University Press (1998).
15. Govind P. Agrawal, *Nonlinear Fiber Optics*, 3rd ed., Academic Press, London (2001).
16. Eugenio Iannone, Francesco Matera, Antonio Mecozzi, and Marina Settembre, *Nonlinear Optical Communication Networks*, Wiley, New York (1998).
17. Yuri S. Kivshar and Govind P. Agrawal, *Optical Solitons*, Academic Press, London (2003).
18. Max Born & Emil Wolf, *Principles of Optics*, Pergamon Press, New York, 1st ed. (1959), 6th ed. (1993).
19. Philip M. Morse and Herman Feshbach, *op. cit.*, one of the classics of the mid 1900s, use the term "tetradic" to denote tensors of rank four.
20. F. W. Peek, Jr., *Dielectric Phenomena in High Voltage Engineering*, 3rd ed., McGraw-Hill, New York (1929).

# Chapter 4
# Vector Calculus Differential Forms
## WITH EXCURSIONS INTO TENSOR CALCULUS

The first step in understanding vector calculus is the comprehension of scalar and vector differential operators—the subject of this chapter. The next step is becoming comfortable with the various integral forms to be able to convert between the differential and integral forms—the subject of the next chapter. Vector differential operators can be made to operate on scalar and vector fields in differing ways yielding other scalar, vector, and dyadic fields. Whereas a scalar differential operator operating on a scalar or vector field will yield another scalar or vector field, respectively, a vector differential operator can yield scalar, vector, or tensor fields depending on its formulative properties and depending upon the tensor rank of the *operand*—the field upon which the operator acts.

A brief introduction to the first-order vector differential operators of gradient, curl, and divergence is given in Section 4.1. In addition, since these operators can be applied to tensors in general, some introductory rules of the gradient, divergence, and curl being applied to tensors are also discussed in this section.

In Section 4.2, scalar differential operators are discussed in general terms, as are ordinary and partial differential equations, eigenvalues, and eigenfunctions. In pointing out that these topics are discussed in general terms, we mean that it is not the intent here to provide a comprehensive study of differential equations. There are many excellent texts that cover these topics quite adequately, such as references 1 through 6, to cite but a few. Besides providing a brief summary of differential equations and their corresponding eigenfunctions, Section 4.2 is intended to paint a picture of scalar differential operators in preparation for an understanding of the directional nature of vector differential operators in contrast with their scalar counterparts.

The *first-order** vector differential operator, known as the *gradient*, is covered in Section 4.3. Two other first-order vector differential operators,

---

* The term *order* is used in mathematics in differing ways. In Chapter 3 we used "order" to refer to the directional compoundedness of a tensor less one. (See Eq. 3.1-1). In this chapter, the "order" of differential operators and differential

namely the *divergence* and *curl*, are covered in Sections 4.4 and 4.5, respectively. Section 4.6 provides some introductory tensor rules for those uninitiated to tensors. These rules are tabulated to show the resulting tensor rank from the application of first-order vector differential operators on tensors of varying rank. *Second-order* differential vector operators include various combinations of the gradient, divergence, and curl, such as the *Laplacian*. These are covered in Section 4.7. In each case of these first- and second-order operators, the operators are described in physical terms and then are expanded in *generalized orthogonal curvilinear coordinates* (GOCCs) with explanations for their use.

Some of the operations are also expanded in *cylindrical coordinates* for two principle reasons. One is that cylindrical coordinates are the simplest of all of the curvilinear orthogonal systems and yet still have properties that require coordinate derivatives of unit vectors to be taken into account. This important point is missed in the usual rush to Cartesian coordinates.

The second and perhaps more important reason for expressing vector operations in cylindrical coordinates is in the *photonics* objective of this book. Photonics includes the vast and highly timely field of optical fibers, which are typically cylindrical in structure and therefore are most naturally analyzed in cylindrical coordinates.

## 4.1 Introduction to Differential Operators
### AND SOME ADDITIONAL TENSOR RULES

As physical sciences began to take a gargantuan leap forward in the mid-1800s, the corresponding mathematical developments took on extreme complexity, often involving ten, twenty, or more coupled, simultaneous, partial differential equations. In these early days before the introduction of the *del* (also called *nabla*) *vector differential operators*, the reduction of large systems of equations to formulate practical solutions to science and engineering questions eluded all but the most proficient scholars. However, with the subsequent development of

---

equations refers to the highest-order derivative in the operator or differential equation, where $\partial^n \Phi / \partial q^n$ is the *n*th-order partial derivative of the function $\Phi(p,q,r)$ with respect to $q$. The order of $\partial^2 \Phi / (\partial p \partial q)$, for example, would be two. Thus, if this derivative were to be part of a differential equation that contained no higher-order derivatives, the order of the differential equation would be ascribed two or *second order*. Likewise, if the differential $\partial^2 / (\partial p \partial q)$ were to be the highest-order differential used in a differential operator, the differential operator would be said to be a *second-order differential operator* or have an order of two.

vector calculus methods that are del-operator based, many of the mysteries of science and the tools of engineering could then be described, understood and applied by the average science or engineering student. These methods, concepts and theorems are reviewed in this and the next chapter.

The glue that has made this somewhat magical transition possible stems, at least in part, from the three basic building block del operators denoted by the inverted, uppercase Greek delta with the truncated name *del*. The first of these, denoted by "$\nabla$" which directly operates* on its operand, is named the *gradient*. The next two are the *del-dot operator*, denoted by "$\nabla\cdot$" and named the *divergence*, and the *del-cross operator*, denoted by "$\nabla\times$" and named the *curl*. Finally, there are various combinations of these first-order operators resulting in higher-order vector differential operators.

As mentioned earlier, another name given to the del operator is "*nabla*." According to Knott,[7] this term was used by Peter Guthrie Tait in personal correspondence with his colleague James Clerk Maxwell in 1870 and perhaps even before then by Robertson Smith who noticed that there was a resemblance of this inverted delta to the shape of an Assyrian harp used by the Hebrews called the *nebel*.**

Before delving into the concepts of gradient, divergence, curl, and their many combinations, it is good for the student to garner an appreciation for the power of these operators to abridge complex mathematical formulations in describing physical phenomena. Further, before examining scalar and vector differential operators, it is beneficial to realize that differential operators can be tensor operators in general. That is, the operator itself can have properties of tensors of any rank. These generalized-rank tensor operators can in turn operate on tensors also having generalized rank. The rules of rank consistency described in Section 3.7 apply for differential operators also. For example, the direct vector del operator, which is the gradient, operating on a tensor of rank $n_R$ will yield a

---

* The term *directly operates* or *direct operator* means that the operation is conducted without the dot- or cross-product type of a process. This is to be used synonymously with *tensor product* for those already familiar with tensors. The tensor product symbol $\otimes$ is used to denote a product between all combinations of the components of each tensor being multiplied resulting in a tensor having a rank equaling the sum of the ranks of the two tensors being multiplied.

** *Webster's Third New International Dictionary* defines *nabla* as "an ancient stringed instrument probably like a Hebrew harp of 10 or 12 strings – also called *nebel*." It is triangular in shape and is held like a harp, thus of the shape of "$\nabla$".

resultant tensor of $n_R + 1$ rank—the same as if a vector were directly multiplying the same $n_R$ rank tensor (without the dot- or cross-product operations). The del dot operator—the divergence—operating on a tensor of rank $n_R$ will yield a new tensor of rank $n_R - 1$. This is also analogous* to the dot product between a vector and a tensor of rank $n_R$, namely, adding the tensor ranks and subtracting two for the inner-product operation (dot product) becomes $1 + n_R - 2 = n_R - 1$.

The rule for the cross-product operation was mentioned in Section 3.8, namely, to subtract one for each cross product from the sum of the ranks of the two tensors. That is, tensor one of rank $n_{R_1}$ crossed (once) with tensor two of rank $n_{R_2}$ results in a new tensor of rank $n_{R_1} + n_{R_2} - 1$. Thus, $[{}_{n_{R_1}}T_1] \times [{}_{n_{R_2}}T_2] = [{}_{(n_{R_1}+n_{R_2}-1)}T_3]$, where the presubscript refers to the rank and the postsubscript distinguishes among the three tensors involved. In particular, if $n_{R_1} = 1$, $[{}_1T_1]$ is a vector and the resultant tensor $[{}_1T_1] \times [{}_{n_{R_2}}T_2] = [{}_{(1+n_{R_2}-1)}T_3] = [{}_{n_{R_2}}T_3]$ has the same rank as $[{}_{n_{R_2}}T_2]$. Further, if the tensor two is also a vector, i.e., $n_{R_1} = n_{R_2} = 1$, then we have the cross product of two vectors yielding a vector resultant, as expected. That is, $[{}_1T_1] \times [{}_1T_2] = [{}_{(1+1-1)}T_3] = [{}_1T_3]$, which in vector notation is $\overline{T}_1 \times \overline{T}_2 = \overline{T}_3$ or $\overline{A} \times \overline{B} = \overline{C}$.

The same rule applies to the curl operator. The curl is a vector operator with an exterior-product-type operation, namely $\nabla \times$. If the curl operates on tensor two $[{}_{n_{R_2}}T_2]$, the result is a new tensor $[{}_{(1+n_{R_2}-1)}T_4]$ having the same rank as $[{}_{n_{R_2}}T_2]$. That is, $\nabla \times [{}_{n_{R_2}}T_2] = [{}_{n_{R_2}}T_4]$. Thus, if the curl acts on a vector, then $n_{R_2} = 1$, and $\nabla \times {}_1T_2 = {}_1T_4$, or $\nabla \times \overline{T}_2 = \overline{T}_4$, or $\nabla \times \overline{A} = \overline{B}$, resulting in another vector as expected.*

Although we will cover the gradient of a vector, which yields a dyadic (in Section 4.3.2), and the divergence of a dyadic, which yields a vector [in Section 4.7.4(a)], we will not delve more deeply into tensorial operations in this chapter. These are developed here because of their importance to subjects like electromagnetics, fiber optics, and materials science (which are often taken in the junior or senior years) and because they serve as practical examples of the

---

* Although the rules of tensor rank consistency apply to the del-dot and del-cross vector operators as they do to the vector dot- and cross-product operations, it is incorrect to apply this reasoning to the operations themselves. As we will see in Sections 4.4 and 4.5, the divergence and curl operations are more involved than just accounting for variations in components of the field upon which it operates, that is, the operand field. Another example of this distinction is in Eq. (4.7-14) and the discussion that follows that equation.

necessity and power of tensor calculus. Instead, we take the position that, except where such operations help in the understanding of the concepts at this level, the main body of this book merely introduces the rules of tensor calculus.[8,9]

## 4.2 Scalar Differential Operators, Differential Equations, and Eigenvalues

Differential operators can be scalar or vector in their nature. For example, a scalar operator $\mathcal{L}$ operating on a scalar field $f(q_1, q_2, q_3, t)$ would be expressed as $\mathcal{L}f$. Here $q_1, q_2, q_3$ are generalized orthogonal coordinates in 3D space and $t$ is time, which is mathematically orthogonal to the spatial coordinates. The homogeneous form of the differential equation utilizing this differential operator is

$$\mathcal{L}f + \lambda f = 0 \tag{4.2-1}$$

where $\lambda$ represents the *eigenvalue*. Although the eigenvalue is not a function of the independent variables $q_1, q_2, q_3, t$, it should not be considered a constant, necessarily. It is a function of the physical parameters and boundary and initial conditions that determine the form of the operator $\mathcal{L}$ and constraints on the values of $f$. For this reason $\lambda$ is often referred to as the *eigenfunction*. Frequently, a great deal can be learned about the solution function $f(q_1, q_2, q_3, t)$ by a detailed study of the eigenfunction. For example, if Eq. (4.2-1) is a wave equation, $\lambda$ describes the *dispersion relation*, which portrays how various components of a signal travel at different velocities and usually degrade the received signal from that which was transmitted. More will be said about this later in this section. First, let us examine some simple scalar operators and their corresponding eigenparameters.

Suppose the scalar function $f(x)$ is a function of a single independent variable $x$, and suppose $\mathcal{L}$ takes the form

$$\mathcal{L} = \frac{d}{dx} \tag{4.2-2}$$

Here $\mathcal{L}$ is a first derivative and is therefore referred to as a first-order scalar differential operator. The differential equation (4.2-1) is then

$$\frac{df}{dx} + \lambda f = 0 \tag{4.2-3}$$

Of course, the solution to Eq. (4.2-3) is of the form

$$f = e^{-ax} \tag{4.2-4}$$

where $a$ is not a function of $x$. The eigenvalue is determined by first differentiating Eq. (4.2-4) and then substituting this into Eq. (4.2-3) yielding $-af + \lambda f = 0$. Since $f$ cannot be zero for all $x$ (otherwise it would be of no use), we may cancel it out, leaving $-a + \lambda = 0$, or

$$\lambda = a \tag{4.2-5}$$

Notice that since $a$ is not a function of $x$, neither is $\lambda$. In fact $\lambda$ represents a damping factor, which is a physical factor of the given problem. Notice also that the above case of a first-order scalar differential operator $L$ operating on the scalar function $f(x)$ yielded a scalar solution of the form $e^{-ax}$.

Next, let us consider a second-order scalar differential operator

$$L = \frac{d^2}{dx^2} \tag{4.2-6}$$

Then the differential equation (4.2-1) becomes

$$\frac{d^2 f}{dx^2} + \lambda f = 0 \tag{4.2-7}$$

Second-order differential equations have two solutions (third-order have three solutions, fourth-order have four solutions, etc.). The solutions to Eq. (4.2-7) take the form

$$f_1 = e^{iax} \tag{4.2-8a}$$

and

$$f_2 = e^{-iax} \tag{4.2-8b}$$

The eigenvalues $\lambda_1$ and $\lambda_2$ are determined by twice differentiating $f_1$ or $f_2$ and solving for $\lambda_1$ and $\lambda_2$, respectively. Since the second derivatives of Eqs. (4.2-8a–b) are $f_1'' = -a^2 f_1$ and $f_2'' = -a^2 f_2$, respectively, $\lambda_1$ and $\lambda_2$ take the same value, namely

$$\lambda_1 = \lambda_2 = a^2 \tag{4.2-9}$$

In such cases $\lambda$ is said to have a *second-order degeneracy*. Notice again that since $a$ is not a function of $x$, neither are the eigenvalues. In fact, since the solution Eqs. [4.2-8(a–b)] are sinusoidal, $\lambda$ represents a periodicity factor, which is a physical factor of the given problem usually associated with the boundary conditions. Notice also that the above case is that of a second-order scalar differential operator $L$ operating on the scalar function $f(x)$ yielding scalar solutions $e^{\pm iax}$.

In many cases, such as in fiber optics, where the fiber core radius varies slightly (either in the manufacturing process or by bending), small perturbations in the solution take place resulting in the slight splitting of the otherwise degenerate eigenvalues. This splitting is referred to as *birefringence*. There are many uses for birefringence in optics and fiber optics including measurements of the order of tens or hundreds of angstroms using interferometer setups where the beat lengths between birefringent signals of the order of millimeters or centimeters can readily be measured. The detailed treatment of second-order degeneracy and birefringence is left for texts in fiber optic communications and nonlinear optics.[10,11]

In the case of our scalar operator $L$ operating on a vector field $\overline{F}(q_1, q_2, q_3, t)$, our result is, of course, a vector field. This is obvious when one expands $\overline{F} = \hat{u}_1 F_1 + \hat{u}_2 F_2 + \hat{u}_3 F_3$, where $\hat{u}_1$, $\hat{u}_2$, and $\hat{u}_3$ are orthogonal unit vectors. Then $L\overline{F} = L\hat{u}_1 F_1 + L\hat{u}_2 F_2 + L\hat{u}_3 F_3$, which is of course vectorial in form. Although tempting, the student is cautioned not to assume that the unit vectors are necessarily constants under the differential operator $L$. In other words, since

$$L\hat{u}_i F_i = \hat{u}_i L F_i + F_i L\hat{u}_i \tag{4.2-10}$$

it is not necessarily correct to assume that $L\hat{u}_i$ is zero. Even though the magnitude of $\hat{u}_i$ is a constant, namely unity, its direction may not be. In spherical coordinates, for example, if $L = \partial/\partial\theta$, then $L\hat{u}_r = \hat{u}_\theta$, which is clearly not equal to zero, and therefore in this case $L\hat{u}_i F_i \neq \hat{u}_i L F_i$. Although $L\hat{u}_i = 0$ in Cartesian coordinates and the longitudinal component of any of the cylindrical coordinate systems, it is not so in general, and both terms of Eq. (4.2-10) must be used.

## 4.3 The Gradient Differential Operator

The *gradient operator* is one of three commonly used first-order vector differential operators. As was pointed out in Section 4.1, its resultant is a tensor of rank one greater than the quantity upon which it operates. Section 4.3.1 provides a physical description of the gradient of a scalar field, derives the gradient from the physical definition,[12] and then expands the gradient into GOCCs. Since the gradient adds one to the rank of the field upon which it operates, the gradient of a scalar field is a vector field.

Likewise, if the gradient operates on a vector field (with a rank of one), the resultant is a tensor of rank two, that is, the resultant is a quantity having a dual directional compoundedness—a dyadic. (See Section 3.7 for rank consistency in equations). One direction is inherent in the vector form of the operator and the other in the vector upon which the operation is being performed. This dual directional compoundedness applies at every point $\bar{r}$ in 3D space and at all times $t$. This dyadic is also presented in GOCCs. Section 4.3.2 thus covers the gradient of a vector field.

### 4.3.1 The gradient of a scalar field—a physical description

Every truck driver knows what it means to enter a grade on the road. The grade can be positive or negative, that is, an uphill grade is considered positive because potential energy increases and a downhill grade is considered negative because the potential energy of the truck decreases. In either case, the driver gears down. In the case of a positive grade, more revolutions of the engine are needed in order to increase the torque on the drive wheels and thus to overcome the increased gravitational force of going uphill—the steeper the hill, the greater is the force and lower is the needed gear. In the case of a negative grade, the driver is even more concerned because the decreasing potential energy is being converted into kinetic energy, which will dangerously increase the speed if action is not taken to absorb that energy. Therefore, the trucker gears down to engage compression braking. In fact, considerable engineering design has gone into designing the compression braking methods to absorb the added energy and prevent it from being converted to its dangerous kinetic form.

Quantitatively, in the definition of the gradient vector differential operator, the direction of the gradient is defined as that of maximal increase of the field upon which it operates. In the above example the field upon which the gradient operates is the scalar gravitational potential energy field. The civil engineer who designs a road to go over a mountain range is, of course, confined to the terrain surface. Although the direction of maximal increase of this potential energy field is straight up, under the constraint of being confined to the terrain surface, the direction of maximal increase is along the steepest slope of the terrain. The engineer knows that vehicles can handle a limited grade. For trucks, that limit is about 6% grade, which means that for every 100 feet of horizontal run the rise is 6 feet or the vertical angle $\theta = \tan^{-1} 0.06 \approx 3.4°$. If the natural slope has places exceeding that limit, the engineer lays out the road at an angle from the direction of steepest ascent thereby limiting the grade of the road to this design limit. The above example is that of a gradient of the gravitational potential energy scalar field on the surface of the earth. Other examples of scalar fields would include temperature, electric potential or voltage, and pressure, to name but a few. Inherent in any of these scalar fields is a gradient. The resultant is a vector field.

Let us next acquire a physical understanding of this vector field that we call the *gradient of a scalar field* from the definition given in Bevc,[11] which was cited as "memorable" by the editor of *American Journal of Physics*, Romer:[13]

> *The gradient of a scalar field is a vector field oriented in the direction in which the scalar field increases most rapidly. Its magnitude is the derivative of the scalar field in the direction of its maximal increase.*

Suppose $V(q_1, q_2, q_3, t)$ is a scalar field. Then the gradient vector $\bar{G}$ of $V$ is by this definition determined by

$$\bar{G} = Grad\ V = \hat{u}_{\max} \lim_{\Delta\ell \to 0} \left. \frac{\Delta V}{\Delta \ell} \right|_{\max} \tag{4.3-1}$$

where $\lim_{\Delta\ell\to 0}(\Delta V/\Delta\ell)_{\max}$ is the maximum rate of increase of $V$ at $P(q_1, q_2, q_3)$ and at time $t$, and where $\hat{u}_{\max}$ is *the* direction of maximal increase. Shorthand notation for the gradient is given by the use of the *del operator* as

$$\bar{G} = \nabla V \tag{4.3-2}$$

First we postulate that the direction of maximal increase is normal to the local surface of constant value of $V$ containing $P$. This postulation is demonstrated in Part (a) below. Next let us construct equivalue surfaces of $V(q_1, q_2, q_3, t)$, using the techniques of Section 2.7, and then construct a unit vector normal to that surface at point $P$ as shown in Fig. 4.3-1. The gradient $\bar{G}$ at point $P(q_1, q_2, q_3)$ and time $t$ of the scalar field $V(q_1, q_2, q_3, t)$ may be written as

$$\bar{G} = \hat{u}_n G_n \tag{4.3-3}$$

where $G_n = \lim_{\Delta\ell \to 0} (\Delta V / \Delta\ell)_{\max}$ and $\hat{u}_n$ is the unit normal at $P(q_1, q_2, q_3)$.

### 4.3.1(a) Why the unit normal *is* the direction of maximal increase

In order to ascertain why the unit normal is the direction of the maximal increase, let us first construct three surfaces each with equal but adjacent values of the scalar field, as shown in Figure 4.3-2. Let the values of the three equivalue surfaces be $V = V_1$, $V_2$, and $V_3$, respectively, such that

$$V_1 < V_2 < V_3 \tag{4.3-4}$$

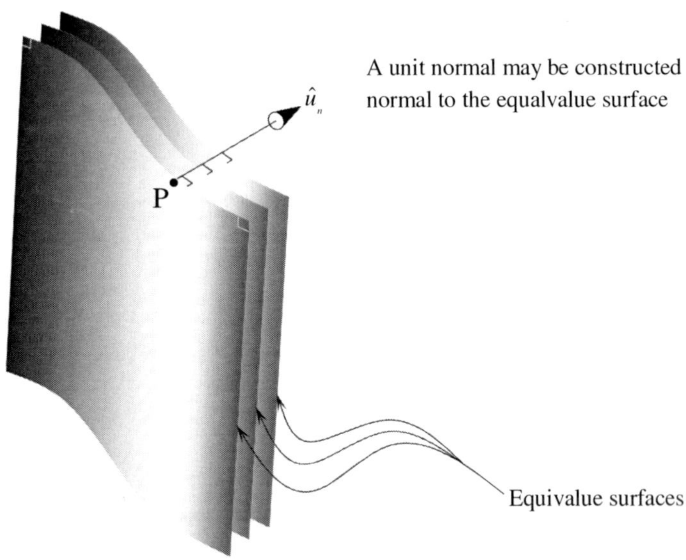


**Figure 4.3-1** Three equivalue surfaces of the scalar field $V(q_1, q_2, q_3, t)$ with a unit normal constructed at a point $P$ on one of the surfaces.

As in Fig. 4.3-1, let us again construct the unit normal vector $\hat{u}_n$, but this time specifically normal to the equivalue surface $V = V_1$ at the point $A$ on $V_1$ as shown in Fig. 4.3-2. Let us construct another unit vector $\hat{u}_1$ also from the point $A$ but not normal to the equivalue surface. Let $\theta$ be the angle between $\hat{u}_1$ and $\hat{u}_n$. Let $\hat{u}_n$ intersect the adjacent surface $V = V_2$ at point $B$ and let $\hat{u}_1$ intersect the same surface $V = V_2$ at point $C$.

Now let us examine the ratio $\Delta V/\Delta \ell$ in Eq. (4.3-1) before taking the limit. The value of the numerator $\Delta V$ from point $A$ on equivalue surface $V = V_1$ to point $B$ on equivalue surface $V = V_2$ is

$$\Delta V_{AB} = V_B - V_A \tag{4.3-5}$$

The value of $\Delta V$ from point $A$ to point $C$ is

$$\Delta V_{AC} = V_C - V_A \tag{4.3-6}$$

Since $V_C$ and $V_B$ are on the same equivalue surface $V_2$, they are equal. Thus,

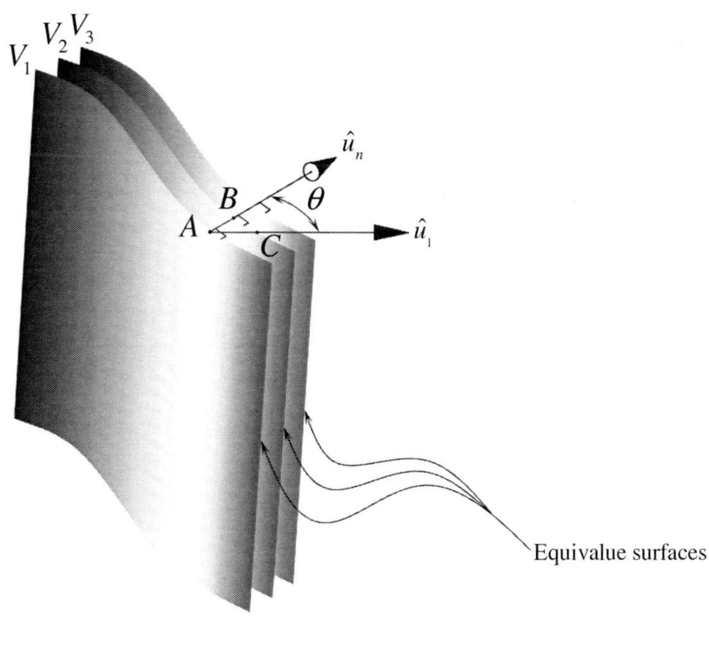


**Figure 4.3-2** Equivalue surfaces of $V = V_1$, $V_2$, and $V_3$ with two unit vectors constructed from a point $A$ on equivalue surface $V = V_1$, where $\hat{u}_n$ is normal to $V_1$ and $\hat{u}_1$ is not.

$$\Delta V = \Delta V_{AB} = \Delta V_{AC} = \Delta V_{12} = V_2 - V_1 > 0 \tag{4.3-7}$$

which is positive because of the condition (4.3-4).

Next, let us look at the denominator $\Delta\ell$. Since

$$\Delta\ell_{AB} = \Delta\ell_{AC}\cos\theta \tag{4.3-8}$$

and since $\theta \neq 0$, $\Delta\ell_{AB} < \Delta\ell_{AC}$ and

$$\left.\frac{\Delta V}{\Delta\ell}\right|_{A \text{ to } B} > \left.\frac{\Delta V}{\Delta\ell}\right|_{A \text{ to } C} \tag{4.3-9}$$

Further, the minimum of $\Delta\ell_{1\text{ to }2}$ occurs at $\theta = 0$, thus $\Delta V/\Delta\ell\big|_n$ is maximal and in passing to the limit, we have

$$\lim_{\Delta l \to 0}\left.\frac{\Delta V}{\Delta\ell}\right|_{\max} = \left.\frac{dV}{d\ell}\right|_n \tag{4.3-10}$$

*Thus, the direction of maximal increase is the direction of the unit normal to the surface of equivalue of* $V(q_1, q_2, q_3, t)$.

### 4.3.1(b) Expansion of the gradient of a scalar field in GOCCs

Before taking the limit in Eq. (4.3-1), we note that

$$G_n \approx \left.\frac{\Delta V}{\Delta\ell}\right|_{\max} \quad \text{for small } \Delta\ell \tag{4.3-11}$$

Multiplying both sides of Eq. (4.3-11) by $\Delta\ell$, we have $\Delta V \approx \Delta\ell_n G_n$. After taking the limit, we have $dV = (\partial V/\partial\ell)d\ell$, where the total differential $d\ell = \overline{d\ell}\big|_n \cdot \hat{u}_n$. Thus, we may write $dV$ as

$$dV = \overline{d\ell}\cdot\overline{G}. \tag{4.3-12}$$

Expanding $\overline{d\ell} = \hat{u}_1 d\ell_1 + \hat{u}_2 d\ell_2 + \hat{u}_3 d\ell_3$, where $d\ell_i = h_i dq_i$,

$$\overline{d\ell} = \hat{u}_1 h_1 dq_1 + \hat{u}_2 h_2 dq_2 + \hat{u}_3 h_3 dq_3 \tag{4.3-13}$$

and

$$\bar{G} = \hat{u}_1 G_1 + \hat{u}_2 G_2 + \hat{u}_3 G_3 \tag{4.3-14}$$

Combining Eqs. (4.3-12), (4.3-13), and (4.3-14) we have

$$dV = G_1 h_1 dq_1 + G_2 h_2 dq_2 + G_3 h_3 dq_3 \tag{4.3-15}$$

Since $V = V(q_1, q_2, q_3)$, a mathematical representation of $dV$ in terms of partial differentials is

$$dV = \frac{\partial V}{\partial \ell} d\ell = \frac{\partial V}{\partial q_1} dq_1 + \frac{\partial V}{\partial q_2} dq_2 + \frac{\partial V}{\partial q_3} dq_3 \tag{4.3-16}$$

Equating coefficients of $dq_i$, we have

$$G_i = \frac{1}{h_i} \frac{\partial V}{\partial q_i} \tag{4.3-17}$$

for each $i$ = 1, 2, 3. Thus, all three components of the gradient $G_i$ may be found from this expression, and therefore

$$\bar{G} = \hat{u}_1 \frac{1}{h_1} \frac{\partial V}{\partial q_1} + \hat{u}_2 \frac{1}{h_2} \frac{\partial V}{\partial q_2} + \hat{u}_3 \frac{1}{h_3} \frac{\partial V}{\partial q_3} \equiv \nabla V \tag{4.3-18}$$

is the expansion of the gradient of a scalar function $V = V(q_1, q_2, q_3)$ in generalized orthogonal curvilinear coordinates.

Whereas equating the coefficients of $dq_i$ in Eqs. (4.3-15) and (4.3-16) in arriving at Eq. (4.3-17) does not demonstrate uniqueness, Bevc does so with elegance in his memorable paper.[12]

### 4.3.1(c) The directional derivative nature of the gradient of a scalar field

Another insight into the physical description of the gradient of a scalar field is given in the concept of the *directional derivative*. Thomas and Finney [Ref. 4, pp 869ff] and Stewart [Ref. 5, pp 756ff] provide mathematical developments of this concept. Although these references develop the directional derivative in rectangular coordinates, we will summarize this in GOCCs. The approach is to find the rate of change of the scalar field (function) $V(q_1, q_2, q_3, t)$ at a point $P(q_1, q_2, q_3)$ in space in the three orthogonal directions $\hat{u}_1, \hat{u}_2$, and $\hat{u}_3$ and obtain the gradient by vectorially adding these three components. Once $\partial V/\partial \ell_i$ is determined in each orthogonal direction $\hat{u}_i$ and after expressing the differential lengths in terms of the metric coefficients $h_i$ and coordinates $\partial q_i$, we have $\partial \ell_i = h_i \partial q_i$ and

$$\nabla V = \sum_{i=1}^{3} \hat{u}_i \frac{1}{h_i} \frac{\partial V}{\partial q_i} \tag{4.3-19}$$

which, of course, is the same as Eq. (4.3-18). By summing the orthogonal derivatives, we have a rate of change of $V(q_1, q_2, q_3, t)$ in *the* direction of maximal increase.

### 4.3.2 The gradient of a vector field

Since the gradient operator does not involve dot- or cross-product operation types (as does the divergence and curl) and since the gradient is *itself* a vector operator, the gradient of a vector field is a dyadic field. That is, the direct product of two tensors, each with rank one, results in a tensor whose rank is the sum of the ranks of the tensors, in this case, two. This is also true for the case of a vector direct-product operator operating on a vector field, i.e., $\overline{BA}$ and $\nabla \overline{A}$ are both dyadics.

#### 4.3.2(a) The gradient of a vector field in GOCCs

Let us determine the gradient of the vector field $\overline{A} = \hat{u}_1 A_1 + \hat{u}_2 A_2 + \hat{u}_3 A_3$. At first glance, the uninitiated might try applying the gradient to each of the scalar components of the vector field $\overline{A}$, and then carry the unit vectors into the vector components of the gradient operator, thus forming nine unit dyads with their appropriate scalar coefficients. However, let us not forget the lesson learned in Eq. (4.2-10), namely that the spatial derivatives of the unit vectors are not zero whenever the direction varies with change in that particular coordinate direction. Therefore, applying Eqs. (4.2-10) and (4.3-16) to each of the vector components of $\overline{A}$, including the unit vectors, we have

$$\begin{aligned} \nabla \overline{A} &= \sum_{i=1}^{3} \sum_{j=i}^{3} \frac{\hat{u}_i}{h_i} \frac{\partial}{\partial q_i} (\hat{u}_j A_j) \\ &= \sum_{i=1}^{3} \sum_{j=1}^{3} \frac{\hat{u}_i \hat{u}_j}{h_i} \frac{\partial A_j}{\partial q_i} + \sum_{i=1}^{3} \sum_{j=1}^{3} \frac{\hat{u}_i A_j}{h_i} \frac{\partial \hat{u}_j}{\partial q_i} \end{aligned} \tag{4.3-20}$$

which is clearly a dyadic. In general, Eq. (4.3-20) contains 18 terms—nine for each of the double summations above. Notice that the second set of nine terms contains all combinations of the three coordinate partial derivatives of each of the three unit vectors. Whereas these nine coordinate derivatives of unit vectors can be expanded with the use of Eqs. (1.3-19) and (1.3-20), we will not do so here because it unnecessarily complicates the efficient form of our dyadic gradient of a vector. Instead, we find it much less cumbersome to perform this expansion once a particular coordinate system is assigned. In either case, the unit vector derivatives on the right may be carried out and the terms collected into components of the resulting nine dyads $\hat{u}_i\hat{u}_j$ or $\hat{\hat{u}}_{ij}$, where $i = 1,2,3$ and $j = 1,2,3$.

### 4.3.2(b) The gradient of a vector field in cylindrical coordinates

In cylindrical coordinates, for example, seven of the nine unit vector derivatives are zero. First, $\hat{u}_z$ does not change direction and all three partial derivatives of $\hat{u}_z$ are therefore zero. Secondly, $\hat{u}_r$ and $\hat{u}_\phi$ do not change direction with changes in $r$ and $z$, thus derivatives with respect to $r$ and $z$ of $\hat{u}_r$ and $\hat{u}_\phi$ are also zero. The only nonzero derivatives are $\partial/\partial\phi$ of $\hat{u}_r$ and $\hat{u}_\phi$. Thus from Eq. (1.3-21)

$$\frac{\partial \hat{u}_r}{\partial \phi} = \hat{u}_\phi \tag{4.3-21}$$

which was derived in Eq.(1.3-22). Further, from Eq. (1.3-20) we have

$$\begin{aligned} \frac{\partial \hat{u}_\phi}{\partial \phi} &= -\frac{\hat{u}_r}{h_r}\frac{\partial h_\phi}{\partial r} = -\frac{\hat{u}_r}{1}\frac{\partial r}{\partial r} \\ \frac{\partial \hat{u}_\phi}{\partial \phi} &= -\hat{u}_r \end{aligned} \tag{4.3-22}$$

Equations (4.3-21) and (4.3-22) are compatible with Ramo, Whinnery, and Van Duzer, (p. 819), for example.[14] Therefore, substituting the subscripts $r, \phi, z$ for $q_1, q_2, q_3$ and $1, r, 1$ for $h_1, h_2, h_3$ in Eq. (4.3-20), we have the expansion of the gradient of the vector field $\overline{A}$ in cylindrical coordinates using Eqs. (4.3-18a and b):

$$\nabla\bar{A}\Big|_{cyl} = +\hat{\hat{u}}_{rr}\frac{\partial A_r}{\partial r}+\hat{\hat{u}}_{r\phi}\frac{\partial A_\phi}{\partial r}+\hat{\hat{u}}_{rz}\frac{\partial A_z}{\partial r}$$
$$+\frac{\hat{\hat{u}}_{\phi r}}{r}\left(\frac{\partial A_r}{\partial \phi}-A_\phi\right)+\frac{\hat{\hat{u}}_{\phi\phi}}{r}\left(\frac{\partial A_\phi}{\partial \phi}+A_r\right)+\frac{\hat{\hat{u}}_{\phi z}}{r}\frac{\partial A_z}{\partial \phi} \tag{4.3-23}$$
$$+\hat{\hat{u}}_{zr}\frac{\partial A_r}{\partial z}+\hat{\hat{u}}_{z\phi}\frac{\partial A_\phi}{\partial z}+\hat{\hat{u}}_{zz}\frac{\partial A_z}{\partial z}$$

which is a dyadic (with its nine unit dyads and their respective scalar coefficients). Notice that two of the nine components, namely the fourth and fifth, each contain two terms stemming from Eqs. (4.3-20) through (4.3-22). The remaining seven do not.

## 4.4 The Divergence Differential Operator

In the introductory paragraph of this chapter, it was stated that vector differential operators can yield scalar, vector, or tensor fields depending on their formulative properties and depending upon the rank of the operand field. In the previous section, we determined that the result of a gradient operation added one to the rank of operand—the gradient of a scalar is a vector, the gradient of a vector is a dyadic, etc. In this section, we review another first-order vector differential operator called the *divergence operator*. The divergence operator is also known as the *del-dot operator*, because it is denoted by the del differential operator followed by dot-product symbol, namely, "$\nabla\cdot$". This operator is entirely different from the gradient operator, even though there are some similarities. They are alike in that both are of first order, both are vectorial in nature, and both use the inverted Greek capital letter delta in their notation. However, they have entirely different properties. In fact, their differences far exceed their similarities. Their inherent definitions are based on entirely different geometries—the gradient on a differential length tends to zero in the limit and the divergence on a differential volume tends to zero in the limit, as we will soon see.

Whereas the divergence operator is not properly formed by simply taking the dot product of the operator with the field that follows, the rules for change in tensor rank do indeed act like the dot product between a vector and the field that follows. That is, the dot (or inner) product, subtracts two from the sum of the ranks. Thus, since the divergence is a vector operator, it has the character of a rank-one tensor. When it operates on a tensor of rank $n_R$, the result is a new tensor of rank $1+n_R-2=n_R-1$ consistent with the rules of Section 3.7. Therefore, the lowest rank tensor that the divergence can operate upon is a vector. From these rules, the divergence of a vector is a scalar and the divergence of a dyadic is a vector, etc.

## 4.4.1 The divergence of a vector field—a physical description

Let us next acquire a physical understanding of this scalar field that results from the *divergence of a vector field* from its definition:

> *The divergence of a vector field at a point in space is the ratio of the net outward* ***flux*** *through an infinitesimal closed surface surrounding the point to the volume enclosed by that surface.*

In mathematical terms, the divergence of the vector field $\overline{A}$ is

$$Div\,\overline{A} \equiv \lim_{\Delta v \to 0} \frac{\oint \overline{A} \cdot \overline{da}}{\Delta v} = \nabla \cdot \overline{A} \qquad (4.4\text{-}1)$$

The numerator represents the net outward flux of $\overline{A}$ through the closed surface and $\Delta v$ is the volume enclosed by the closed surface. This is graphically depicted in Fig. 4.4-1.

Before attempting to convey a physical understanding of the concept of divergence, we need to first garner the concept of the flux of a vector field. Every vector field can be thought of as a field of fluxes. The total flux passing through a surface $S$ is determinable by taking the dot product of the vector field at every differential element of area on $S$ with the vector differential surface area $\overline{da}$ and integrating over $S$. Thus, the flux $\Psi_A$ of the vector field $\overline{A}$ through the surface $S$ is

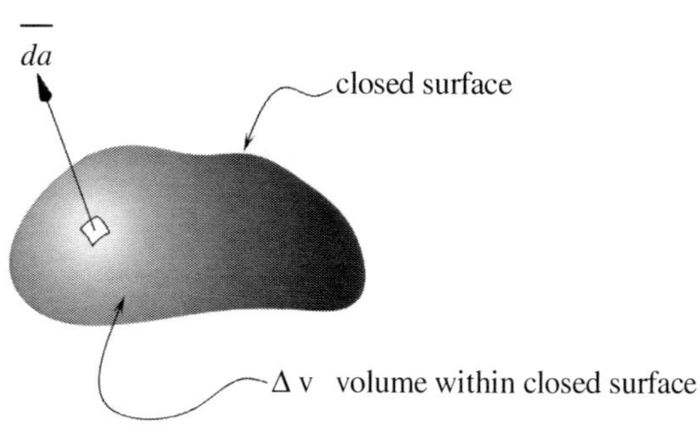


**Figure 4.4-1** The geometry associated with the definition of divergence.

$$\Psi_A = \int_S \overline{A} \cdot \overline{da} \tag{4.4-2}$$

It is worthwhile to note that since $\overline{da}$ is everywhere perpendicular to $S$, any portion of the surface $S$ that is parallel with the field $\overline{A}$ makes no contribution to the flux because of the dot product.

Further, whenever the surface $S$ is a completely closed surface, the integral represents the total outward flux given by

$$\Psi_{A_{out}} = \oint \overline{A} \cdot \overline{da} \tag{4.4-3}$$

Therefore the numerator of Eq. (4.4-1) represents the total outward flux.

### 4.4.1(a) Vector-field flux tubes and sources

Vector fields exist because of a source. For example, a gravitational force field exists because of the presence of mass. A static electric field exists because of the presence of charge. A fluid flow field exists because an external drive, such as a pump, a fan, or a turbine causes fluid (either gas or liquid) to move. A static magnetic field exists because of the presence of a steady current or because of the presence of a magnetized object. In every case, the vector field is the effect of some external source, and in every case, the field exists beyond the boundaries of the source. That is, these vector fields are present at points in space for which there are no sources as long as at least one source at another position in space exists to cause the field in the first place. The shape of the field depends on the placement and distribution of the sources, the properties of the medium hosting the field, and the boundary constraints or *boundary conditions*.

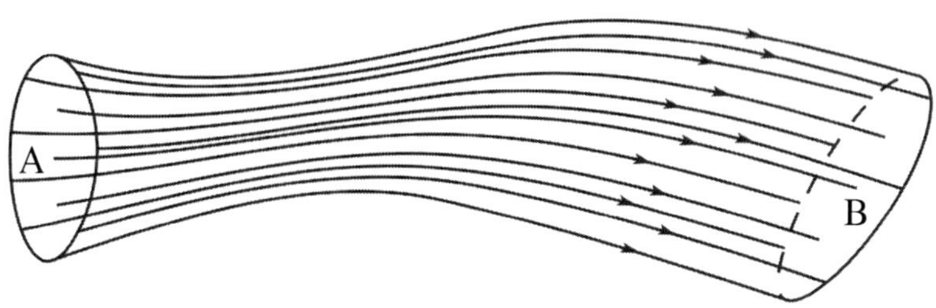


**Figure 4.4-2** Graphical representation of a generalized vector field flux tube with nine field direction lines defining the side of the tube and three field direction lines shown in the interior of the flux tube.

Consider a given vector field $\overline{F}$. Applying the techniques of Section 2.6, vector field direction lines (also called flow lines) can be graphed. Consider collecting an imaginary bundle of these flow lines into a tube whose sides are everywhere parallel to the flow lines truncated at ends by two surfaces $A$ and $B$ that intersect the bundle of flow lines not necessarily at right angles to them. Such a tube can be thought of as a field of flux flow lines or a *flux tube*. Figure 4.4-2 provides a pictorial representation of a vector field flux tube.

Flux tubes have the characteristic that no flux penetrates the sides of the tube. Consider a volume to be made up of a section of a flux tube as shown. Therefore, the only flux entering or leaving the volume are through the cross-sectional end caps of the flux tube. If the same amount of flux enters the volume through end-cap $A$ that leaves through end-cap $B$, then by the definition above, the divergence is said to be zero even though the field direction lines may be physically diverging or converging. It is possible for the divergence of a given vector field $\overline{F}$ not to be zero, that is, more flux enters through end cap $A$ than leaves through end cap $B$, or vice versa. This can happen only if there is a distribution of sources in the volume of the tube, as we will observe.

### 4.4.1(b) Examples of zero and nonzero divergence

In order to provide an understanding of divergence from its definition above, we will postulate some simple vector fields and derive the divergence from their corresponding flux tubes without the use of a coordinate system. Of these examples, some will have zero and some nonzero divergence despite whether the flux lines physically diverge. In Part (c), we will discuss the significance of the nonzero divergence on the distribution of sources within the volume of the closed surface.

#### *Example (1a):*

Given a vector field that is of constant magnitude everywhere but directed outward from a point $P$ in space,

$$\overline{A} = k\hat{u}_R \tag{4.4-4}$$

where $k$ is a constant and $\hat{u}_r$ is a unit vector directed away from $P$. Next let us construct a flux tube that consists of a conical section of arbitrary cross-sectional shape and of end caps at $R = R_1$ and $R = R_2$, where $R$ is measured from $P$, as shown in Fig. 4.4-3 and where $R_1$ and $R_2$ are constants with $R_1 < R_2$.

The end caps are surfaces of spheres having radii $R_1$ and $R_2$. The areas of the two surfaces at $R_1$ and $R_2$ are $S_1 = \Omega R_1^2$ and $S_2 = \Omega R_2^2$, respectively, where $\Omega$ is the solid angle subtended by the generalized cone (expressed in units of steradians). Since the field is uniform throughout from Eq. (4.4-4), the flux entering the $R = R_1$ surface is $\Psi_1 = k\Omega R_1^2$ and the flux leaving the $R = R_2$ surface is $\Psi_2 = k\Omega R_2^2$. Since no flux penetrates the conical surface on the sides, the total outward flux is $k\Omega(R_2^2 - R_1^2)$.

The volume of the closed surface is $\frac{1}{3}\Omega(R_2^3 - R_1^3)$. Before passing to the limit in Eq. (4.4-1), we can examine the ratio of the outward flux to the volume, specifically

$$\frac{k\Omega(R_2^2 - R_1^2)}{\frac{1}{3}\Omega(R_2^3 - R_1^3)} = \frac{3k(R_2^2 - R_1^2)}{(R_2^3 - R_1^3)}. \tag{4.4-5}$$

In anticipation of passing to the limit as $\Delta v \to 0$, we need to let $R_2$ approach $R_1$. This can be accomplished by letting $R_2 = R_1 + \Delta R$ and solving Eq. (4.4-5) as $\Delta R \to 0$. Normally in the limit as $\Delta v \to 0$ we would have to simultaneously let both $\Delta R \to 0$ as well as $\Omega \to 0$. But since $\Omega$ dropped out of our ratio, it is not necessary to impose the limit $\Omega \to 0$.

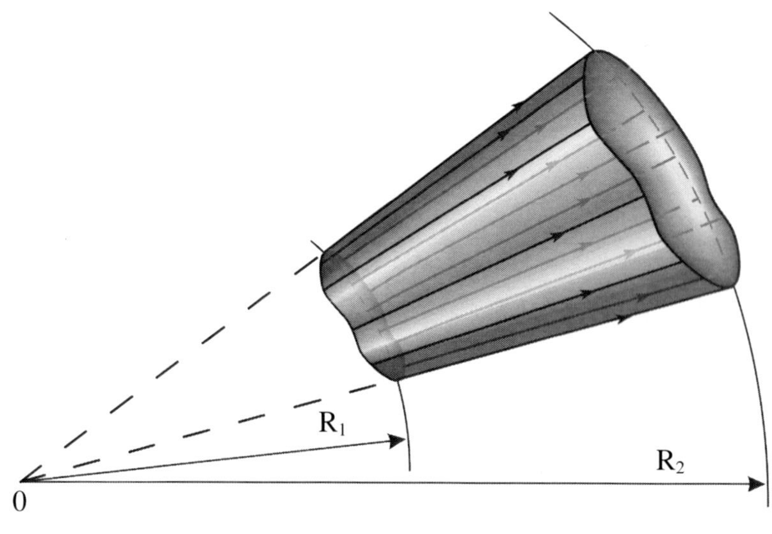


**Figure 4.4-3** Closed-surface flux tube for Eqs. (4.4-4), (4.4-7) and (4.4-9).

Let us expand the numerator and denominator of Eq. (4.4-5) in a Taylor series. The numerator becomes $R_2^2 - R_1^2 = 2R_1\Delta R + O(\Delta R)^2$ and the denominator becomes $R_2^3 - R_1^3 = 3R_1^2\Delta R + O(\Delta R)^2$, where $O(\Delta R)^2$ refers to a term *of the order of* $(\Delta R)^2$. In passing to the limit as $\Delta R \to 0$, the terms $O(\Delta R)^2$ are negligible with respect to the terms that vary as $\Delta R$. The ratio, then, is the divergence of Eq. (4.4-4), namely

$$\nabla \cdot \overline{A} = \frac{2k}{R} \tag{4.4-6}$$

which was determined for Example (1a) from the definition without the use of a coordinate system.

*Example (1b):*

Next, let us consider another vector field

$$\overline{A} = kR\hat{u}_R \tag{4.4-7}$$

The flux tube for this field is also represented by Fig. 4.4-3. However, the field in this case grows in magnitude uniformly with $R$. The areas of the two surfaces at $R_1$ and $R_2$ are unchanged, being $S_1 = \Omega R_1^2$ and $S_2 = \Omega R_2^2$; however, the flux entering the $R = R_1$ surface is $\overline{A}(R_1)\cdot S_1\hat{u}_R$, and therefore $\Psi_1 = kR_1\Omega R_1^2 = k\Omega R_1^3$. Likewise, the flux leaving the $R = R_2$ surface is $\Psi_2 = k\Omega R_2^3$, and the total outward flux is $k\Omega(R_2^3 - R_1^3)$. The volume is the same as in the previous example giving us the resultant divergence of Eq. (4.4-5) as

$$\nabla \cdot \overline{A} = \frac{k\Omega(R_2^3 - R_1^3)}{\frac{1}{3}\Omega(R_2^3 - R_1^3)} = 3k \tag{4.4-8}$$

In this case, the ratio became a constant without actually having to pass to the limit of infinitesimal $\Delta R$.

*Example (1c):*

Let us generalize the prior two examples by letting the vector fields be

$$\overline{A} = kR^p\hat{u}_R \tag{4.4-9}$$

where $p$ is any general power. Note that $p = 0$ and $1$ for Examples (1a) and (1b), respectively. Again, we will find the divergence from the definition.

Again, the flux tube for this field is represented by Fig. 4.4-3. However, the field in this case changes in magnitude uniformly with $R^p$. The areas of the two surfaces at $R_1$ and $R_2$ are again unchanged and the flux entering at $R_1$ and leaving at $R_2$ are $\Psi_{1,2} = k\Omega R_{1,2}^{p+2}$, respectively. The total outward flux is $k\Omega(R_2^{p+2} - R_1^{p+2})$. The volume is unchanged from the previous two examples.

As in the first example, we let $R_2 = R_1 + \Delta R$ and expand $R_2$ in a Taylor series. $\Delta R$ then takes the form $R_2^3 - R_1^3 = (p+2)R_1^{p+1}\Delta R + O(\Delta R)^2$. As before, the volume expanded in a Taylor series is $R_2^3 - R_1^3 = 3R_1^2\Delta R + O(\Delta R)^2$, yielding the resultant divergence of Eq. (4.4-9) as $\Delta R \to 0$ as

$$\nabla \cdot \overline{A} = k(p+2)R^{p-1} \tag{4.4-10}$$

Notice that Eq. (4.4-10) reduces to $2k/R$ for $p = 0$, consistent with Eq. (4.4-6) and to $3k$ for $p = 1$, consistent with Eq. (4.4-8).

More importantly, however, is the case of $p = -2$. Here the vector field is

$$\overline{A} = \frac{k}{R^2}\hat{u}_r \tag{4.4-11}$$

From Eq. (4.4-10), the divergence is therefore

$$\nabla \cdot \overline{A} = 0 \tag{4.4-12}$$

This result is also evident when one realizes that the surface increases as $R^2$, whereas the field drops off as $R^{-2}$. Therefore, the flux entering the end cap $S_1$ is equal to the flux leaving the end cap $S_2$ and therefore the numerator is always zero. There are many examples of vector fields that behave as Eq. (4.4-11). The gravitational field in the space external to a mass falls off as $R^{-2}$, where $R$ is the distance from the center of gravity of the mass. The electric flux density external to a charge falls off as $R^{-2}$, where $R$ is the distance from the charge. Each of these fields has a zero divergence even though the field direction lines appear to be diverging.

### Example (2a):

In examples (1a), (1b), and (1c), vector fields were mathematically created to be physically diverging from a point, whereas the divergence was positive, negative, or even zero. Therefore, whether a field has a divergence is not necessarily related to whether it in fact diverges (or converges) physically. To further illustrate this lack of connection between appearance of the field lines and the existence of a divergence value, we will create two additional examples of fields that have parallel field lines, that is, direction field lines that are not converging or diverging. One of these will have a nonzero divergence, the other will have a zero value.

Suppose $\overline{A} = x\hat{u}_x$. For our closed surface, let us create a cube in the region $1 \le x, y, z \le 2$. The field flow lines of $\overline{A}$ are parallel to four of the six surfaces, namely the two that are parallel to the *x*-*y* plane and the two that are parallel to the *x*-*z* plane. The total outward flux is calculated as the flux leaving the surface at $x = 2$ less the flux entering at $x = 1$, which is $2 - 1 = 1$. Therefore, $\nabla \cdot \overline{A} = 1$, even though the flux lines are not diverging physically.

### Example (2b):

Again, let us pick a field where all flux lines are not physically diverging. Suppose $\overline{A} = y\hat{u}_x$ and, again, let us reconstruct the same cube. The flux lines are parallel to the same four sides and penetrate the same two sides as before. In this case, however, the flux entering at $x = 1$ is $1\frac{1}{2}$ and the flux leaving at $x = 2$ is also $1\frac{1}{2}$. Thus, total outward flux is $1\frac{1}{2} - 1\frac{1}{2} = 0$, and $\nabla \cdot \overline{A} = 0$.

In these last two examples, neither field was physically diverging or converging, yet one was zero and one nonzero. Therefore, whether there is an appearance of divergence or convergence in the field flow lines is uncorrelated with whether the divergence of the field is zero.

## 4.4.1(c) Significance of a nonzero divergence

In the prior examples we found no relationship between the physical appearance of vector flux lines diverging and the value of the divergence of vector the field. So, we ask the following:

*On what does the divergence depend?*

The answer is *sources*. That is, if the divergence is nonzero in any region in space, there must be a distribution of sources in that region. In particular, the divergence is equal to the volume density of the source. For example, in any region where there is a distribution of mass, such as in the interior of the Earth, there is a mass density throughout the volume of that region. Thus, at every point in the interior of the earth there is a mass density in kg/m$^3$. This mass density is the source of the divergence, and is, in fact, equal to the divergence.

In the case of electric flux density $\overline{D}$, its source is electric charge density $\rho$ in coulombs per meter cubed (C/m$^3$). From the definition of divergence, Eq. (4.4-1), the divergence is the limit of the ratio of the net outward flux through an infinitesimal closed surface to the volume enclosed by that surface as the volume goes to zero. The total charge enclosed within the infinitesimal closed surface is the volume integral

$$Q_{total} = \int_{\Delta v} \rho dv \qquad (4.4\text{-}13)$$

### 4.4.2 The divergence in GOCCs

The expansion of the divergence of a vector field $\overline{A} = \hat{u}_1 A_1 + \hat{u}_2 A_2 + \hat{u}_3 A_3$ in generalized orthogonal curvilinear coordinates is accomplished by carefully adding the net outward flux through the six orthogonal differential surfaces that bound an infinitesimal volume. The sides lie on surfaces of constant value of each of the coordinates in three pairs each separated by differential amounts. The three pairs are front-back, right-left, and top-bottom. This volume is pictorially represented in Figure 4.4-4.

The total outward flux of $\overline{A}$ is accomplished by summing the following six integrals

$$\begin{aligned} \Psi_{\overline{A}} = \oint \overline{A} \cdot \overline{da} = & \int_{front} + \int_{back} \\ & + \int_{right} + \int_{left} \\ & + \int_{top} + \int_{bottom} \end{aligned} \qquad (4.4\text{-}14)$$

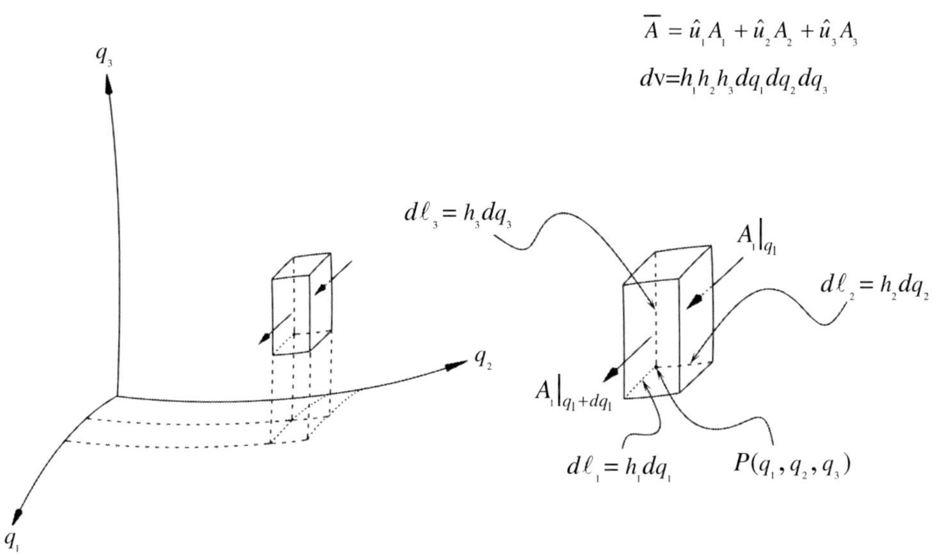


**Figure 4.4-4** Volume element used in $\nabla \cdot \overline{A}$ derivation in GOCCs.

In performing these integrations it is necessary to account for variations in surface elements as well as the vector components when one changes a coordinate. We will first examine the first pair. The *front* integral is over an elemental surface at $q_1 + \Delta q_1$ having an area of $\Delta\ell_2\Delta\ell_3$ and in the $\hat{u}_1$ direction. Therefore, the vector representation for the differential area at the *front* element is $\overline{da}_1$. Here the subscript denotes the vectorial direction of the elemental surface, $\overline{da}_1 = \hat{u}_1\Delta\ell_2\Delta\ell_3\Big|_{q_1+\Delta q_1}$. Similarly, the flux through the *front* surface is

$$\Psi_{A_1 front} = \int_{front} A_1\hat{u}_1 \cdot \overbrace{\hat{u}_1 da_1}^{\overline{da}_1}\Big|_{q_1+\Delta q_1} = A_1\Delta\ell_2\Delta\ell_3\Big|_{q_1+\Delta q_1} \qquad (4.4\text{-}15)$$

To complete the first pair, we note that the *back* integral is over an elemental surface located at $q_1$, except that the outward direction is $-\hat{u}_1$. Thus, we label the vector differential area as $\overline{da}_{-1}$. With these changes, the procedure for finding the outward flux through the *back* surface is otherwise the same, yielding

$$\Psi_{A_1 back} = \int_{back} A_1\hat{u}_1 \cdot \overbrace{(-\hat{u}_1) da_1}^{\overline{da}_{-1}}\Big|_{q_1} = A_1\Delta\ell_2\Delta\ell_3\Big|_{q_1} \qquad (4.4\text{-}16)$$

The net of the flux *front* and *back* is

$$\Psi_{A_1}\Big|_{front+back} = A_1\Delta\ell_2\Delta\ell_3\Big|_{q_1+\Delta q_1} - A_1\Delta\ell_2\Delta\ell_3\Big|_{q_1} \tag{4.4-17}$$

Variations in the surface element $\Delta\ell_2\Delta\ell_3$ as well as variations in the vector component $A_1$ from $q_1$ to $q_1+\Delta q_1$ must both be taken into account with respect to the differential lengths $\Delta\ell_1$, $\Delta\ell_2$, and $\Delta\ell_3$. This is expected for the component but the variations in the surface elements may be overlooked by the less mathematically discerning. Length variations are related to coordinate variations by $\Delta\ell_i = h_i\Delta q_i$ from Eq. (2.6-5), where the metric coefficients in general are functions of the coordinates, i.e., $h_i = h_i(q_1, q_2, q_3)$.

Before passing to the limit, we note that the differential volume is given by Eqs. (1.2-6) and (2.6-5)

$$\Delta v = h_1h_2h_3\Delta q_1\Delta q_2\Delta q_3. \tag{4.4-18}$$

Therefore,

$$\lim_{\Delta v\to 0}\frac{1}{\Delta v}\left[\int_{front}+\int_{back}\right] = \frac{h_2h_3\Delta q_2\Delta q_3A_1\Big|_{q_1+\Delta q_1} - h_2h_3\Delta q_2\Delta q_3A_1\Big|_{q_1}}{h_1h_2h_3\Delta q_1\Delta q_2\Delta q_3} \tag{4.4-19}$$

As stated before, it is necessary to account for variations in surface elements, such as $h_2h_3\Delta q_2\Delta q_3$, as well as the vector components, which in this case is $A_1$, as we vary from $q_1$ to $q_1+\Delta q_1$. Since there is no variation in coordinates $q_2$ and $q_3$ with respect to $q_1$, because they are orthogonal, the necessity to account for variations in surface elements leaves only variations in the metric coefficients, $h_2$ and $h_3$, which *are*, in general, functions of coordinates, as pointed out above. Therefore, the $\Delta q_2\Delta q_3$ factors in Eq. (4.4-18) cancel in the numerator and denominator, leaving variations in the component and metric coefficients to be considered. Thus, Eq. (4.4-18) becomes

$$\lim_{\Delta v\to 0}\frac{1}{\Delta v}\left[\int_{front}+\int_{back}\right] = \frac{1}{h_1h_2h_3}\lim_{\Delta q_1\to 0}\frac{h_2h_3A_1\big|_{q_1+\Delta q_1} - h_2h_3A_1\big|_{q_1}}{\Delta q_1} \tag{4.4-20}$$

Noting that the right-hand factor is precisely the definition of the partial derivative, we have

$$\lim_{\Delta v \to 0} \frac{1}{\Delta \text{v}} \left[ \int_{front} + \int_{back} \right] = \frac{1}{h_1 h_2 h_3} \frac{\partial (h_2 h_3 A_1)}{\partial q_1} \qquad (4.4\text{-}21)$$

for one of the three scalar terms making up the divergence of the vector field $\overline{A}$.

The same process may be repeated for the other two pairs of integrals; however, this tedious procedure is not necessary because we may simply *roll the subscripts** to obtain the remaining two terms of the divergence. The *roll* sequence is $1 \to 2 \to 3 \to 1 \to 2$. Therefore, the divergence of our vector field $\overline{A}$ is

$$\begin{aligned} \nabla \cdot \overline{A} &= \frac{1}{h_1 h_2 h_3} \left[ \frac{\partial (h_2 h_3 A_1)}{\partial q_1} + \frac{\partial (h_3 h_1 A_2)}{\partial q_2} + \frac{\partial (h_1 h_2 A_3)}{\partial q_3} \right] \\ &= \frac{1}{h_1 h_2 h_3} \sum_{i=1}^{3} \frac{\partial}{\partial q_i} \left( \frac{h_1 h_2 h_3}{h_i} A_i \right) \end{aligned} \qquad (4.4\text{-}22)$$

This equation is specialized for Cartesian coordinates in Appendix B, Eq. (B.1-3) and for cylindrical coordinates in Eq. (B.3-9).

## 4.5 The Curl Differential Operator

The *curl operator* is the third of the three first-order vector differential operators introduced in Section 4.1. Whereas the gradient employed the del operator ($\nabla$) directly and the divergence employed the del-dot operator ($\nabla \cdot$), the curl employs the *del-cross operator*, denoted by "$\nabla \times$". In the previous section, we found that divergence of a vector could not in general be found by simply taking the dot product of the del operator with the vector because it was necessary to account for variations in surface elements as well as the vector components. Here we will find a similar admonition. The curl of a vector field is not simply the cross product of the del operator with the vector for a similar reason. Although one can validly get by with this misleading approach when expanding the curl in

---

* One of the paramount advantages of using generalized coordinates (GOCCs) is the ability to roll subscripts. When expanding vector operators into GOCCs in 3D space, it is necessary to do so for one third of the problem. The remaining two-thirds may be deduced by simply rolling the subscripts. This procedure is invalid in all specific (nongeneralized) coordinate systems except Cartesian coordinates. In Cartesian coordinates, it is permissible because all three metric coefficients, $h_x, h_y, h_z$, are unity and thus do not have variations with respect to coordinate directions. In this special case, the roll sequence is $x \to y \to z \to x \to y$.

Cartesian coordinates, it is invalid to do so in any other coordinate system. Many texts that first introduce the student to vector differential operators immediately fall into this oversimplistic approach of expanding these operators in rectangular coordinates, no doubt because to immediately expand into GOCCs exposes the student to an inordinate complexity of calculative rules before providing a perception of the nature of the operator. In the case of the curl, Bevc states this with his usual insight and precision:[12]

> *To be sure, such rules are useful in actual calculations but they hardly provide any physical insight into the nature of the curl and moreover depend on coordinate systems.*

In this section we find that the curl does not change the rank of the field upon which it operates. This is in contrast to the previous two operators in that the result of a gradient operation added one to the rank and the divergence subtracted one from the rank of field that was operated upon. Therefore, if the three operators act on a vector field, the divergence will yield a scalar, the curl will yield a vector, and the gradient will yield a dyadic.

Like the gradient and the divergence, the curl is a first-order vector operator using the del notation; however, the similarities end there. The curl operator is entirely different from the two just previously reviewed. The inherent definitions are based on three entirely different geometries—the gradient on a differential length going to zero in the limit, the curl on a differential area going to zero in the limit (as we will soon see), and the divergence on a differential volume going to zero in the limit. In addition, *the curl operates transversely, whereas the divergence operates tangentially and the gradient operates omni-versely, so-to-speak.* By this, we do not mean that the resultant is transverse, tangential, etc.—just the operand acts in these ways.

### 4.5.1 The curl of a vector field—a physical description

Let us next acquire a physical understanding of this vector field that we call the *curl of a vector field* from the definition (outside of the context of coordinates). Again, the definition of this vector operation is given by Bevc[12] (with emphasis added):

> *The curl of a vector field $\overline{A}$ at a point is a vector pointing in the direction of a normal to an infinitesimal surface which is so oriented in space that the limit of the ratio of the line integral of the vector*

*field $\overline{A}$ around the perimeter of that surface to the area enclosed is* ***maximal****. The magnitude of the curl is the value of that limit.*

Mathematically, the curl vector $\overline{C}$ of the vector field $\overline{A}$ is by this definition determined by

$$\overline{C} = curl\ \overline{A} \equiv \lim_{\Delta a \to 0} \left. \frac{\oint \overline{A} \cdot \overline{d\ell}}{\Delta a} \right|_{\max} \hat{u}_n \tag{4.5-1}$$

where

$$\lim_{\Delta a \to 0} \left( \text{circ}(\overline{A}) / \Delta a \right) \Big|_{\max}$$

is the maximum of the ratio of the *circulation of* $\overline{A}(\overline{r},t)$, defined by Eq. (2.4-22), about the point $P$ located at $\overline{r}$ in space and at time $t$ to the enclosed area, and where $\hat{u}_n$ is the normal to that surface at $P$ in the right-hand sense with respect to the direction of the closed-line integration. Shorthand notation for the curl is given by the use of the *del-cross operator* as

$$\overline{C} = \nabla \times \overline{A} \tag{4.5-2}$$

Since there are a triply infinite number of closed paths about a point—an infinite number of paths about each of the three orthogonal axes passing through the point $P$—it may seem that finding the maximal ratio would be a formidable task. However, a perfectly straightforward procedure is taken to resolve this difficulty.

First, a component of the curl in an arbitrary direction, say $\hat{u}_1$, is found from the above definition. That is, an arbitrarily selected infinitesimal surface, $\Delta a_1$, is chosen with $\hat{u}_1$ as its normal. This surface is planar and is bounded by an infinitesimal closed path $\overline{d\ell}$, whose direction is taken in the right-hand sense (that is, with the thumb of the right hand in the direction of $\hat{u}_1$, the fingers give the direction of the closed path). It is chosen such that the plane of the path contains the point $P$ at which the curl of $\overline{A}$ is desired. In performing the limit, namely $\lim_{\Delta a \to 0} \left( \text{circ}(\overline{A}) / \Delta a \right)$, the vector component of the curl of $\overline{A}$ in the $\hat{u}_1$ direction is determined:

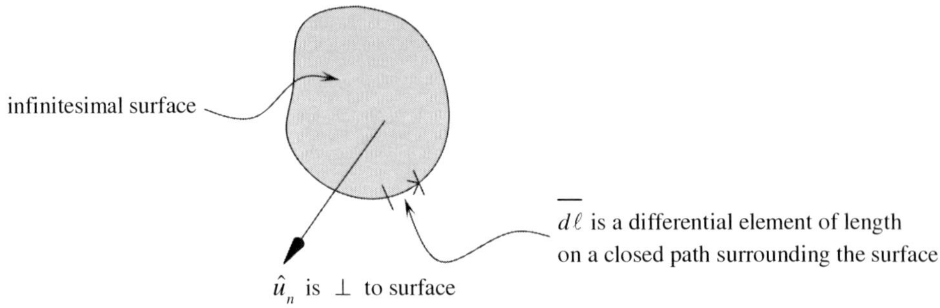


**Figure 4.5-1** The geometry associated with the definition of curl.

$$curl\,\overline{A}\Big|_1 = \hat{u}_1 \lim_{\Delta a_1 \to 0} \frac{\oint_{\Delta a_1} \overline{A}\cdot\overline{d\ell}}{\Delta a_1} \tag{4.5-3}$$

A second direction, say $\hat{u}_2$, is taken orthogonal to the first but otherwise arbitrary, and the procedure is repeated to obtain the second component. Finally, a third component is taken orthogonal to the first two by the right-hand rule (and, thus, is uniquely determined). We assign its direction as $\hat{u}_3$, and repeat the procedure again. Summing the three orthogonal components, the resulting vector *is* the desired maximal ratio and is the curl of $\overline{A}$:

$$\begin{aligned} \nabla\times\overline{A} &= \sum_{i=1}^{3} curl\,\mathrm{A}\Big|_i \\ &= \hat{u}_i \sum_{i=1}^{3} \lim_{\Delta a_i \to 0} \frac{\oint_{\Delta a_i} \mathrm{A}\cdot\overline{d\ell}}{\Delta a_i} \end{aligned} \tag{4.5-4}$$

Note that this expression was determined from the definition without the need for any coordinate system.

### 4.5.2 The curl as a vorticity vector

In order to give further physical interpretation of the curl operator we need to garner a physical understanding of the *circulation integral*, Eq. (4.5-1) — an intimate part of the definition of the curl. As first presented in Eq. (2.4-22), the circulation of the vector field $\overline{A}$ is

$$circ(\overline{\mathrm{A}}) = \oint \overline{A}\cdot\overline{d\ell} \tag{4.5-5}$$

From [Morse & Feshbach, pp 18ff][15] *this integral is a measure of the tendency of the field's flow lines to "curl up."* In cases such as magnetic fields or fluid flow fields where the field direction lines either close on themselves or circulate as in a helix, the circulation of the field, $circ(\overline{A})$, will not be zero. As defined in the discussion following Eq. (2.4-22), such fields are referred to as *rotational, solenoidal,* or *nonconservative.* Other terms expressing this circulatory nature of some fields are *paddle-wheeling*[4,5] (Thomas & Finney, p. 992 and Schwarz, p. 154ff), *swirl*[2] (O'Neil, p. 972), and *vorticity*[16] (Rogers, p. 275). Each of these terms conjures up the image of circulating or twirling fields.

The *paddle-wheel* concept is perhaps the easiest to understand for the student's initial exposure to curl. Suppose that a small paddle wheel consisting of symmetrical, uniform, planar fins on an axial shaft is placed in a fluid that is flowing. If the flow lines are uniform, that is, having constant direction and strength, the paddle wheel will not rotate no matter what the direction of its axis is. However, if there is a variation in the flow field, either in magnitude or direction or both, there will be orientations of the axis in which the paddle wheel will rotate. The rotational speed of the paddle wheel is a measure of the magnitude of the vector component of the curl. The axis is the direction of the component, where the thumb of the right hand gives the orientation of the direction when the fingers are orientated in the direction of rotation. As the axis is adjusted for maximum rotation, the ultimate curl vector is empirically determined. This postulation may be tested by rotating the axis in each of two orthogonal directions and noting that the paddle wheel does not turn in either of these orientations. Thus, the component of the curl that exhibits maximum circulation where the other two orthogonal components are zero *is* the curl.

Such a *gedanken experiment* (German for "thought experiment") may be tested by the construction of a *curl meter,* which consists of a small paddle wheel metered to display its angular velocity. As with most such instruments, the presence of the probe may affect the field that it measures; however, the instrument can often be oriented to minimize such errors.

The curl operator is a measure of the *circulation density* or *vorticity* of a vector field[15]—that is, the circulation per unit cross-sectional area—which is precisely given in the definition of the curl, Eq. (4.5-1). As Morse & Feshbach point out, the limiting process of Eq. (4.5-1) "is more complicated than that used to define the divergence, for the results obtained depend on the orientation of the element of area," another way of pointing out the ultimate task of determining the *maximal* ratio specified by the definition. In their ensuing discussion Morse &

Feshbach elegantly demonstrate the validity of this task by simply determining the three orthogonal vector components of the curl, as was done in Section 4.5.1.

### 4.5.3 The expansion of the curl in GOCCs

The expansion of the curl of a vector field $\overline{A} = \hat{u}_1 A_1 + \hat{u}_2 A_2 + \hat{u}_3 A_3$ in generalized orthogonal curvilinear coordinates is derived from the physical definition[11] by carefully accounting for the circulation density about three orthogonal surfaces $\overline{da_1}$, $\overline{da_2}$, and $\overline{da_3}$ in each orthogonal direction in the generalized system $q_1, q_2, q_3$. The first of these surfaces, namely $\overline{da_1}$, is depicted in Figure 4.5-2 in order to determine the first component of the curl, given by

$$(\nabla \times \overline{A})_1 = \hat{u}_1 \lim_{\Delta a_1 \to 0} \frac{\oint \overline{A} \cdot \overline{d\ell}}{\Delta a_1} \tag{4.5-6}$$

The circulation of $\overline{da_1}$ is carefully determined while accounting for variations in the differential lengths as well as vector components while changing coordinates. To accentuate these variations, we will again represent the differential lengths as $\Delta\ell_i$ and, thus, the circulation integral around surface $\overline{da_1}$ will involve the sum of four integrals—first from 1 to 2, then from 2 to 3, then 3 to 4, and finally from 4 back to 1. The integral from 1 to 2 is taken along $\Delta\ell_2$ at $q_3$, while the integral from 3 to 4 is taken along $-\Delta\ell_2$ at $q_3 + \Delta q_3$. The integral from 2 to 3 is taken along $\Delta\ell_3$ at $q_2 + \Delta q_2$, and the integral from 4 to 1 is taken along $-\Delta\ell_3$ at $q_2$.

The circulation integral in Eq. (4.5-6) is the sum of four integrals as described in the paragraph above. Therefore, the first component of the curl may be written as

$$(\nabla \times \overline{A})_1 = \hat{u}_1 \lim_{\Delta a_1 \to 0} \frac{\int_1^2 + \int_2^3 + \int_3^4 + \int_4^1}{\Delta a_1} \tag{4.5-7}$$

Since the first and third integrals are taken along $\Delta\ell_2$ and $-\Delta\ell_2$ at $q_3$ and $q_3 + \Delta q_3$, respectively, and the second and fourth integrals are taken along $\Delta\ell_3$ and $-\Delta\ell_3$ at $q_2 + \Delta q_2$ and $q_2$, respectively, we organize these into two separate limits as follows:

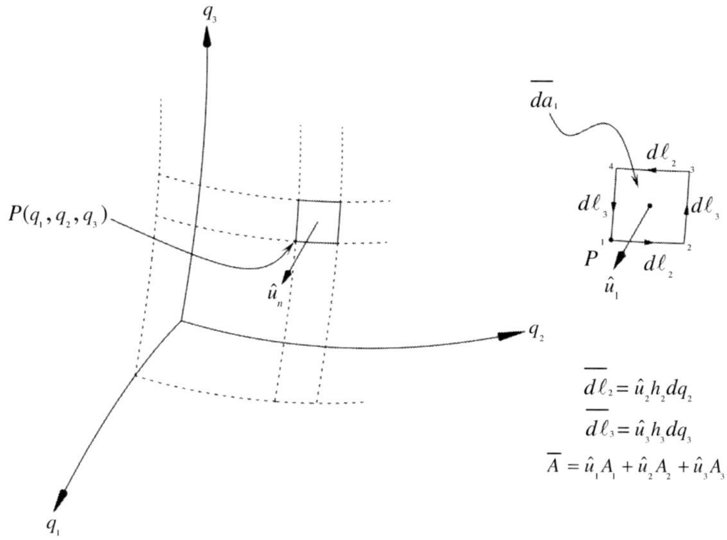


**Figure 4.5-2** Surface element use in the derivation of the first component of the curl $(\nabla \times \overline{A})_1$ in GOCCs.

$$(\nabla \times \overline{A})_1 = \hat{u}_1 \lim_{\Delta a_1 \to 0} \frac{\int_1^2 + \int_3^4}{\Delta a_1} + \hat{u}_1 \lim_{\Delta a_1 \to 0} \frac{\int_2^3 + \int_4^1}{\Delta a_1} \tag{4.5-8}$$

where $\Delta a_1 = \Delta \ell_2 \Delta \ell_3 = h_2 h_3 \Delta q_2 \Delta q_3$. Since the first limit above is taken over variations in $q_3$ and the second limit above is taken over variations in $q_2$, limits occur first in Eq. (4.5-8) as $\Delta q_3 \to 0$ and then as $\Delta q_2 \to 0$. Thus, component one of the curl is found by the following process:

$$\begin{aligned}(\nabla \times \overline{A})_1 &= \hat{u}_1 \lim_{\Delta q_3 \to 0} \frac{\overbrace{\left[A_2 \hat{u}_2 \cdot \hat{u}_2 h_2\right]_{q_3}}^{\int_1^2} \Delta q_2 + \overbrace{\left[A_2 \hat{u}_2 \cdot (-\hat{u}_2) h_2\right]_{q_3 + \Delta q_3}}^{\int_3^4} \Delta q_2}{h_2 h_3 \Delta q_2 \Delta q_3} \\ &+ \hat{u}_1 \lim_{\Delta q_2 \to 0} \frac{\overbrace{\left[A_3 \hat{u}_3 \cdot (-\hat{u}_3) h_3\right]_{q_2}}^{\int_4^1} \Delta q_3 + \overbrace{\left[A_3 \hat{u}_3 \cdot \hat{u}_3 h_3\right]_{q_2 + \Delta q_2}}^{\int_2^3} \Delta q_3}{h_2 h_3 \Delta q_2 \Delta q_3}\end{aligned} \tag{4.5-9}$$

where the variations in lengths reduce to variations in metric coefficients. Since the coordinates are orthogonal, there are no variations in coordinates $q_2$ and $q_3$, and therefore these may be cancelled in the numerators and denominators. Thus, we have

$$(\nabla\times\overline{A})_1=\frac{\hat{u}_1}{h_2h_3}\left(\lim_{\Delta q_3\to 0}\frac{\left[A_2h_2\right]_{q_3}-\left[A_2h_2\right]_{q_3+\Delta q_3}}{\Delta q_3}+\lim_{\Delta q_2\to 0}\frac{-\left[A_3h_3\right]_{q_2}+\left[A_3h_3\right]_{q_2+\Delta q_2}}{\Delta q_2}\right) \tag{4.5-10}$$

which are readily recogonized as partial derivatives. Therefore, the first component of the curl of $\overline{A}$ is

$$(\nabla\times\overline{A})_1=\frac{\hat{u}_1}{h_2h_3}\left[\frac{\partial\ (A_3h_3)}{\partial q_2}-\frac{\partial(A_2h_2)}{\partial q_3}\right] \tag{4.5-11}$$

This is the component of the curl in the $\hat{u}_1$ direction. The next two components are determined in the same manner; however, these may be written out simply by "rolling" the subscripts. Summing the components, we have

$$\nabla\times\overline{A}=\frac{1}{h_1h_2h_3}\sum_{i=1}^{3}\hat{u}_ih_i\left[\frac{\partial}{\partial q_j}(h_kA_k)-\frac{\partial}{\partial q_k}(h_jA_j)\right] \tag{4.5-12a}$$

where $j=i+1=2,3,1$ and $k=j+1=3,1,2$. Another convenient form (often preferred by students) for the curl of a vector is

$$\nabla\times\overline{A}=\begin{vmatrix}\dfrac{\hat{u}_1}{h_2h_3} & \dfrac{\hat{u}_2}{h_3h_1} & \dfrac{\hat{u}_3}{h_1h_2}\\ \dfrac{\partial}{\partial q_1} & \dfrac{\partial}{\partial q_2} & \dfrac{\partial}{\partial q_3}\\ h_1A_1 & h_2A_2 & h_3A_3\end{vmatrix} \tag{4.5-12b}$$

which can be readily expanded to arrive at Eq. (4.5-12a) after expanding the summation.

### 4.5.4 The expansion of the curl in cylindrical coordinates

Substituting $r, \phi, z$ for $q_1, q_2, q_3$ and $1, r, 1$ for $h_1, h_2, h_3$ in Eq. (4.5-12a) we have

$$\nabla\times\bar{A}\Big|_{cyl} = \hat{u}_r\left[\frac{1}{r}\frac{\partial A_z}{\partial\phi} - \frac{\partial A_\phi}{\partial z}\right] + \hat{u}_\phi\left[\frac{\partial A_r}{\partial z} - \frac{\partial A_z}{\partial r}\right] + \hat{u}_z\frac{1}{r}\left[\frac{\partial(rA_\phi)}{\partial r} - \frac{\partial A_r}{\partial\phi}\right] \quad (4.5\text{-}13a)$$

or alternatively in determinant form, we have

$$\nabla\times\bar{A}\Big|_{cyl} = \begin{vmatrix} \dfrac{\hat{u}_r}{r} & \hat{u}_\phi & \dfrac{\hat{u}_z}{r} \\ \dfrac{\partial}{\partial r} & \dfrac{\partial}{\partial\phi} & \dfrac{\partial}{\partial z} \\ A_r & rA_\phi & A_z \end{vmatrix} \quad (4.5\text{-}13b)$$

## 4.6 Tensorial Resultants of First-Order Vector Differential Operators

To summarize, let us tabulate the resultant quantities from the three first-order vector differential operators developed in the preceding three sections. We will first establish single-character symbols—**D**, **C** and **G**—to denote divergence, curl and gradient, respectively. This ordering is chosen in increasing order of resultant tensor rank. That is, the divergence, curl, and gradient change the rank of the operand—the quantity upon which they operate—by $-1, 0, +1$, respectively. As stated in Section 4.1, a vector differential operator can yield scalar, vector, or tensor fields depending on its properties and depending upon the rank of the operand. Table 4-1 summarizes, encapsulates, and generalizes this statement for the divergence, curl, and gradient of scalars, vectors, dyadics and tensors in general.

Since there can be no quantity with negative rank, the divergence cannot operate on a scalar. Also, by careful inspection of Eq. (4.5-12a), the curl cannot operate on a scalar either. These observations are consistent with the rules for the dot and cross products between vectors. One cannot take a dot or cross product of a vector with a scalar. For the same reason, one cannot take the divergence or

**Table 4-1** Resultant tensor rank from three first-order vector differential operators.

| Diff. Vector Operator | with a scalar $(n_R = 0)$ operand **s** | with a vector $(n_R = 1)$ operand **v** | with a dyadic $(n_R = 2)$ operand **d** | with a tensor of rank $n_R$ operand ${}_{n_R}T$ |
|---|---|---|---|---|
| **D** | x | **s** $(n_R = 0)$ | **v** $(n_R = 1)$ | ${}_{n_R-1}T$ |
| **C** | x | **v** $(n_R = 1)$ | **d** $(n_R = 2)$ | ${}_{n_R}T$ |
| **G** | **v** $(n_R = 1)$ | **d** $(n_R = 2)$ | **t** $(n_R = 3)$ | ${}_{n_R+1}T$ |

Key:

| | | |
|---|---|---|
| **x** ⇒ nonexistent | **d** ⇒ *diadic* $(n_R = 2)$ | **D**=divergence |
| **s** ⇒ *scalar* $(n_R = 0)$ | **t** ⇒ *triadic* $(n_R = 3)$ | **C**=curl |
| **v** ⇒ *vector* $(n_R = 1)$ | ${}_{n_R}T$ ⇒ tensor of rank $n_R$ | **G**=gradient |

curl* of a scalar. Therefore, these two operations are noted as "nonexistent" in Table 4-1.

Note that resultant quantities align diagonally in Table 4-1. For example, the operations **Gs**, **Cv**, and **Dd** result in vectors, which line up diagonally. Likewise, **Gv**, **Cd**, and **Dt** (the latter, **D** operating on a triadic, tensor of rank $n_R = 3$) also line up diagonally, each having dyadic resultants.

## 4.7 Second-Order Vector Differential Operators—Differential Operators of Differential Operators

Thus far we have been dealing with the three classical first-order vector differential operators—divergence, curl, and gradient. In this section, we will cover some of the combinations of these. There are logically nine combinations of these three operators, although some may be nonexistent and some may be zero depending upon the quantity being operated on, which we call the *operand*. Again, in order to list and sort these nine combinations, let us use the same single-character symbols, namely **D**, **C**, and **G**, that we use in building the above table, to denote divergence, curl, and gradient, respectively. This ordering is

* Although the curl of a scalar is considered nonexistent, if such an operation did exist in some sense—a pure abstraction—it would be a scalar, since the curl does not change the rank of the quantity upon which it operates.

chosen because *resultant* tensor rank from these operators changes in upward steps, namely $-1, 0, +1$, respectively, as previously pointed out.

In Section 4.7.1 the various combinations of second-order vector differential operators and operands are tabulated in terms of their respective resultant forms. This exercise in tabulating the nature of the resultant forms from second-order operations before studying the details of these operations is intended to provide focus to the understanding of the operations and garner an appreciation for their significant features.

The subsequent sections provide detailed explanations of six of the commonly used second-order combinations. In Section 4.7.2 two combinations that involve the curl become zero. These are shown to have considerable significance in formulating real-world solutions to abstract physical phenomena using the tools of vector calculus, such as Maxwell's equations in electromagnetics and Schrödinger's equations in quantum physics and nonlinear optics. Sections 4.7.3 and 4.7.4 cover combinations leading to the scalar and vector Laplacian operators and 4.7.5 and 4.7.6 round off our detailing of the commonly used second-order operators.

### 4.7.1 Resultant forms from second-order vector differential operators—a tabular summary of tensorial resultants

The nine combinations of second-order differential operations taken in **D**,**C**,**G** order would be **DD**, **DC**, **DG**, **CD**, **CC**, **CG**, **GD**, **GC**, and **GG**, each operating on scalars, vectors, dyadics, and general rank tensor operands. Because there are nine resultant quantities potentially for each of the four operands, we will create four tables. Each table lists "first operator" in columns and "second operator" in rows, where first and second refer to the steps that one takes in performing the differential operations. For example, "**DGs**," which denotes the divergence of the gradient of a scalar, such as $\nabla \cdot \nabla V$, is performed by first executing the **Gs** operation. That is, the gradient of the scalar, for our example $\nabla V$, is performed first (as it must, because the divergence could not be done first). Tables 4-2(a), (b), (c), and (d) list the nine resultant operations on scalars (**s**), vectors (**v**), dyadics (**d**), and general rank tensor ${}_{n_R}T$ operands, respectively. Since there are nine combinations, we give the designation **PQ** to denote any one of the nine.

*The key at the end of the tables should be used in order to garner their full impact and to follow the explanations of the information.*

In our example above **DGs** is be found on Table 4-2(a) (because the operand **s** is a scalar), in the third column (because the first operation **Gs** is the gradient **G**), and in the first row, (because the second operation is the divergence **D**). The resultant of **DGs** is a scalar listed as "**s**" on Table 4-2(a), **G** column, **D** row. Further, since **DGs** is a common operation (called the Laplacian of a scalar), a reference is given to Section 4.7.3 where **DGs** is covered.

The principal features of each of these tables are described in the five paragraphs that follow.

In Table 4-2(a) six of the nine second-order operators are nonexistent since neither the divergence nor the curl can have scalar operands. Only the gradient can. Thus, **DG**, **CG**, and **GG** yield scalar, vector, and dyadic resultants, respectively. **DGs** is the *scalar Laplacian* and is discussed in Section 4.7.3 and **CGs** is one of the operations involving the curl that becomes identically zero as shown in Section 4.7.2. These section references are indicated in the table. **GGs** has a dyadic resultant; however, it is not generally used in upper-division engineering and physical science courses. Therefore, we give no further details in this guide.

In Table 4-2(b) two of the nine combinations are nonexistent because the divergence of a vector is a scalar and a scalar cannot be an operand of divergence or curl. Four of the nine combinations have referrals to subsequent sections. **DCv** refers to Section 4.7.2, since it is another important second-order operator that becomes zero. In addition **CGv** is zero, which means that all nine components of the dyadic are zero.* **DGv**, which is the *vector Laplacian*, is referenced to Section 4.7.4. **CCv** and **GDv** are referenced to Sections 4.7.5, and 4.7.6, respectively.

Table 4.2(c) provides the resultants for the nine second-order differential operators with dyadic operands. All nine resultants exist (or are zero) starting with a scalar in the upper left, two vectors on the next diagonal, then three dyadics on the next diagonal, followed by two triadics, and finally by one quadadic in the lower right. However, two of these—**DCd** and **CGd**—are zero, meaning that the three components of the vector resultant of **DCd** and the 27 components of the triadic resultant of **CGd** are zero.*

---

* The fact that second-order differential operators **DC** and **CG** are zero for any operand regardless of rank is demonstrated (at an intermediate tensor calculus level) in Appendix C, Sections C.2-2 and C.2-3, respectively.

**Table 4-2** Second-order vector differential operator resultant forms with four operands—scalars, vectors, dyadics and generalized tensors.

(a) Scalar operand **s** ($n_R = 0$)

| **PQ**s<br>Second operation **P** (1st symbol) ↓ / First operation **Q** (2nd symbol) → | **D** | **C** | **G** |
|---|---|---|---|
| **D** | x | x | **s** [4.7.3] |
| **C** | x | x | **v**=0* [4.7.2] |
| **G** | x | x | **d** ($n_R = 2$) |

(b) Vector operand **v** ($n_R = 1$)

| **PQ**v<br>Second operation **P** (1st symbol) ↓ / First operation **Q** (2nd symbol) → | **D** | **C** | **G** |
|---|---|---|---|
| **D** | x | **s**=0* [4.7.2] | **v** [4.7.4] |
| **C** | x | **v** [4.7.5] | **d**=0* ($n_R = 2$) |
| **G** | **v** [4.7.6] | **d** ($n_R = 2$) | **t** ($n_R = 3$) |

(c) Dyadic operand **d** ($n_R = 2$)

| **PQ**d<br>Second operation **P** (1st symbol) ↓ / First operation **Q** (2nd symbol) → | **D** | **C** | **G** |
|---|---|---|---|
| **D** | **s** | **v**=0* | **d** ($n_R = 2$) |
| **C** | **v** | **d** ($n_R = 2$) | **t**=0* ($n_R = 3$) |
| **G** | **d** ($n_R = 2$) | **t** ($n_R = 3$) | **q** ($n_R = 4$) |

(d) Tensor operand generalized rank: $n_R \geq 2$

| **PQ** $n_R$<br>Second operation **P** (1st symbol) ↓ / First operation **Q** (2nd symbol) → | **D** | **C** | **G** |
|---|---|---|---|
| **D** | ${}_{n_R-2}T$ | ${}_{x}T = 0^*$ $x = n_R - 1$ | ${}_{n_R}T$ |
| **C** | ${}_{n_R-1}T$ | ${}_{n_R}T$ | ${}_{x}T = 0^*$ $x = n_R + 1$ * |
| **G** | ${}_{n_R}T$ | ${}_{n_R+1}T$ | ${}_{n_R+2}T$ |

Key:

**x** $\Rightarrow$ *nonexistent*
**s** $\Rightarrow$ *scalar* ($n_R = 0$)
**v** $\Rightarrow$ *vector* ($n_R = 1$)
**d** $\Rightarrow$ *diadic* ($n_R = 2$)

**t** $\Rightarrow$ *triadic* ($n_R = 3$)
**q** $\Rightarrow$ *quadadic* ($n_R = 4$)
$n_R$ $\Rightarrow$ *tensor rank*
${}_{n_R}T$ $\Rightarrow$ *tensor of rank* $n_R$

* ${}_{n_R}T = 0 \Rightarrow$ *all tensor components are zero*

**D**=divergence
**C**=curl
**G**=gradient
**P**=**D**,**C** or **G**
**Q**=**D**,**C** or **G**

Table 4.2(d) generalizes the first three for all existent cases, including those that are zero. All nine second-order operators are assumed to operate on a tensor ${}_{n_R}T$ (of rank $n_R$) called the operand. The upper-left resultant is a tensor ${}_{n_R-2}T$ (of rank $n_R-2$). The two resultants in the next diagonal are tensors ${}_{n_R-1}T$ (of rank $n_R-1$), although all components of the divergence of the curl turn out to be zero. The three resultants in the next diagonal are different tensors, but of the same rank as the operand, and are different from their operand, in general. On the next diagonal, the two resultant tensors ${}_{n_R+1}T$ are of rank $n_R+1$; however, again, one has components that are all zero, namely the curl of the gradient. Finally, the gradient of the gradient (lower right corner) yields a resultant ${}_{n_R+2}T$ tensor (of rank $n_R+2$).

Of the 36 combinations of three operators taken two at a time with four operands, eight are non-nonexistent, six are detailed in the subsequent section because of their importance to juniors and seniors, and the remaining cite only their respective resultants for further study. Of these 36 combinations, seven are identically zero, meaning that all tensor components of these seven are zero. These are denoted by the asterisks in Table 4-2.

### 4.7.2 Two important second-order vector differential operators that vanish

As stated above and referenced in Tables 4-2(a) and (b), **CGs** and **DCv** become zero. We will demonstrate these two identities and discuss the significance of their vanishing, which is more than the casual observer might expect. First, **CGs** $=\nabla\times\nabla V$. Combining Eqs. (4.5-12a) and (4.3-19), we have

$$\nabla\times\nabla V=\frac{1}{h_1h_2h_3}\sum_{i=1}^{3}\hat{u}_ih_i\left[\frac{\partial}{\partial q_{i+1}}\left(h_{i+2}\frac{1}{h_{i+2}}\frac{\partial V}{\partial q_{i+2}}\right)-\frac{\partial}{\partial q_{i+2}}\left(h_{i+1}\frac{1}{h_{i+1}}\frac{\partial V}{\partial q_{i+1}}\right)\right]$$

Notice that the metric coefficients $h_{i+1}$ and $h_{i+2}$ cancel. Also canceling are the resulting second derivatives, which from Eq. (1.3-11) are independent of the differentiation order, that is, $\partial^2/(\partial q_{i+1}\partial q_{i+2})=\partial^2/(\partial q_{i+2}\partial q_{i+1})$. Thus, the contents of the square brackets vanish and we have a vector identity

$$\nabla\times\nabla V=0 \tag{4.7-1}$$

Therefore, since **CGs** = 0 from Eq. (4.7-1), we show this result in Table 4-2(a), column **G**, row **C** as **v**=0, since the gradient of a scalar is a vector and in turn the curl of that vector is another vector. This means that all three components of **v** are identically zero.

*Significance:* This says that anytime we have a conservative vector field—a field whose curl is zero—that field may be represented by the gradient of a scalar. For example, Maxwell's curl equation for the electric field intensity is zero in electrostatics, meaning that the electric field intensity is conservative: $\nabla \times \overline{E} = 0$. The implication of Eq. (4.7-1) is that the electric field intensity may be written as $\overline{E} = -\nabla V$. (The minus sign originates from the sign of the charge of the electron). In this example, $V$ is the electric scalar potential (volts in SI units) and $\overline{E}$ is the (vector) electric field intensity (volts per meter in SI units).

Next let us explore **DCv** $= \nabla \cdot \nabla \times \overline{A}$. The divergence of the curl of a vector field can be found by combining Eqs. (4.4-22) and (4.5-12a):

$$\nabla \cdot \nabla \times \overline{A} = \frac{1}{h_1 h_2 h_3} \left( \begin{aligned} &\frac{\partial}{\partial q_1}\left\{ h_2 h_3 \frac{1}{h_2 h_3}\left[ \frac{\partial\,(A_3 h_3)}{\partial q_2} - \frac{\partial (A_2 h_2)}{\partial q_3} \right]\right\} \\ &+\frac{\partial}{\partial q_2}\left\{ h_3 h_1 \frac{1}{h_3 h_1}\left[ \frac{\partial\,(A_1 h_1)}{\partial q_3} - \frac{\partial (A_3 h_3)}{\partial q_1} \right]\right\} \\ &+\frac{\partial}{\partial q_3}\left\{ h_1 h_2 \frac{1}{h_1 h_2}\left[ \frac{\partial\,(A_2 h_2)}{\partial q_1} - \frac{\partial (A_1 h_1)}{\partial q_2} \right]\right\} \end{aligned} \right) \tag{4.7-2}$$

Notice again that the metric coefficient combinations cancel and the resulting six second derivatives, which from the mixed derivative theorem Eq. (1.3-11) are independent of the differentiation order, also cancel. Thus, the contents of the large parentheses vanish and we have the vector identity

$$\nabla \cdot \nabla \times \overline{A} = 0 \tag{4.7-3}$$

Therefore, since **DCv** = 0 from Eq. (4.7-3), we display this result in Table 4-2(b), column **C**, row **D** as **s** = 0, since the curl of a vector is a vector, and the divergence of that vector is a scalar.

*Significance*: This shows that anytime we have a vector field with a zero divergence, that vector field can be represented by the curl of another vector field. For example, Maxwell's equation for the magnetic flux density is $\nabla \cdot \overline{B} = 0$ in electromagnetics, which says that there are no magnetic monopoles in physics.

Therefore, from Eq. (4.7-3), the magnetic flux density vector field $\overline{B}$ may be written as $\overline{B} = \nabla \times \overline{A}$. In this example, $\overline{A}$ is the magnetic vector potential.

### 4.7.3 The divergence of the gradient of a scalar field—the scalar Laplacian

One of the more commonly used second-order vector differential operators is the divergence of the gradient of a scalar field ($\nabla \cdot \nabla V$). The result is a scalar, as one can see by first taking the gradient of the scalar field $V(q_1, q_2, q_3)$, which is a vector, and in turn taking the divergence of that vector yielding a scalar.

#### 4.7.3(a) The scalar Laplacian in GOCCs

Using Eq. (4.3-18), we determined the gradient of $V$ in GOCCs. Substituting this result for the vector $\overline{A}$ in Eq. (4.4-22), one obtains

$$\nabla \cdot \nabla V = \frac{1}{h_1 h_2 h_3}\left[\frac{\partial}{\partial q_1}\left(\frac{h_2 h_3}{h_1}\frac{\partial V}{\partial q_1}\right) + \frac{\partial}{\partial q_2}\left(\frac{h_3 h_1}{h_2}\frac{\partial V}{\partial q_2}\right) + \frac{\partial}{\partial q_3}\left(\frac{h_1 h_2}{h_3}\frac{\partial V}{\partial q_3}\right)\right] \tag{4.7-4}$$

The detailed development of the above expression is left as an exercise for the student. There being only scalars and vectors involved, it should be a perfectly straightforward process for students at the junior or senior level.

This second-order vector differential operation is used so frequently in physics and engineering that a special symbol is used to simplify the notation, like so:

$$\nabla \cdot \nabla V \equiv \nabla^2 V \tag{4.7-5}$$

Note that the shorthand notation for the divergence of the gradient operator **DG** $= \nabla \cdot \nabla = \nabla^2$, where $\nabla^2$ is called the *del-squared operator*. This operator was first introduced by Maxwell as *Laplace's operator*. In modern parlance, we call it the *Laplacian*. We have already pointed out that the Laplacian of a scalar (**DGs**) is a scalar and noted it in Table 4.2(a), column **G**, row **D**. Moreover, since both the resultant and the operand are scalars, **DGs** = $\nabla^2 V$ is also called the *scalar Laplacian*.

It is noteworthy that the Laplacian does not change the rank of the operand in general. This is evident from column **G**, row **D** in Tables 4-2(b), (c), and (d), each of which has a resultant with the same rank as its operand. In Section 4.7.4 we will deal with another extremely important and frequently used case, namely, the Laplacian of a vector. First, however, we will express Eq. (4.7-4) in cylindrical coordinates.

### 4.7.3(b) The scalar Laplacian in cylindrical coordinates

Substituting $r, \phi, z$ for $q_1, q_2, q_3$ and $1, r, 1$ for $h_1, h_2, h_3$ in Eq. (4.7-4) we have

$$\nabla^2 V\Big|_{cyl} = \nabla \cdot \nabla V = \frac{1}{r}\frac{\partial}{\partial r}\left(r\frac{\partial V}{\partial r}\right) + \frac{1}{r^2}\frac{\partial^2 V}{\partial \phi^2} + \frac{\partial^2 V}{\partial z^2} \qquad (4.7\text{-}6)$$

which is the cylindrical coordinate expansion of the scalar Laplacian.

## 4.7.4 The divergence of the gradient of a vector field—the vector Laplacian

Perhaps equally as common as the scalar Laplacian, if not more so, is the *vector Laplacian.* This second-order vector differential operator is the divergence of the gradient of a vector field (**DGv** $= \nabla \cdot \nabla \overline{A}$). The result is a vector, as was noted earlier. There is, however, a crucial difference in its implementation compared with the scalar Laplacian. One can see by first taking the gradient of the vector field $\overline{A}(q_1, q_2, q_3)$ that the resultant is a dyadic. The case of the gradient of a vector field was developed in Section 4.3.2 and resulted in the dyadic given by Eq. (4.3-20). However, we cannot simply substitute this into Eq. (4.4-22), since we are dealing here with the divergence of a dyadic and Eq. (4.4-22) already has had the inner-product operation on a vector built into it. The resulting scalar form of Eq. (4.4-22) has no provision for the insertion of the nine dyadic components.

Therefore, in Part (a) of this section, we develop the divergence of a dyadic. The vector Laplacian is built upon this result, first in GOCCs [Part (b)], and, then, to illustrate the process of having to take the spatial derivatives of the unit vectors, expanded in cylindrical coordinates in Part (c).

### 4.7.4(a) The divergence of a dyadic in GOCCs

Since the inner-product operation involved in the divergence of a vector was already implemented in the development of Eq. (4.4-22) resulting in a scalar, it is of little use to us in setting up the operator in a form that can be applied to tensors of higher rank. Let us first look at the nature of the divergence operator in such a form that it may be applied to any tensor operand:

$$\nabla\cdot\ \ = \frac{1}{h_1h_2h_3}\sum_{i=1}^{3}\frac{\partial}{\partial q_i}\left(h_{i+1}h_{i+2}\hat{u}_i\cdot\ \ \right) \qquad (4.7\text{-}7)$$

where the blank spaces after the dots are left for insertion of the tensor components, including the unit tensors in each component. Notice the position of the unit vector $\hat{u}_i$ and the position of the dot operator. Since the unit vector in general can vary with changes in coordinates, it would be improper to place the unit vector to the left of the derivative. This statement can be tested by treating the divergence of a vector operand and checking whether Eq. (4.7-7) reduces to Eq. (4.4-22). Let $\overline{A} = \hat{u}_1A_1 + \hat{u}_2A_2 + \hat{u}_3A_3$ and insert this into Eq. (4.7-7). One can readily show that this results in Eq. (4.4-22). However, if Eq. (4.7-7) were written in the form $\sum\hat{u}_i\,\partial\left(h_{i+1}h_{i+2}\cdot\hat{u}_iA_i\right)/\partial q_i$ or $\sum\hat{u}_i\cdot\left[\partial\left(h_{i+1}h_{i+2}\hat{u}_iA_i\right)/\partial q_i\right]$, the derivative would have to be taken on the internal unit vector and the resultant scalar would not be Eq. (4.4-22).

Next, to determine the divergence of a dyadic, consider the dyadic $\overline{\overline{G}}$ in generalized coordinates

$$\overline{\overline{G}} = \sum_{i=1}^{3}\sum_{j=1}^{3}\hat{u}_i\hat{u}_jG_{ij} \qquad (4.7\text{-}8)$$

Notice the dual directional compoundedness of the unit dyads $\hat{u}_i\hat{u}_j$ and the nine scalar components $G_{ij}$. Substituting this into Eq. (4.7-7), we have

$$\nabla\cdot\overline{\overline{G}} = \frac{1}{h_1h_2h_3}\sum_{i=1}^{3}\sum_{j=1}^{3}\frac{\partial}{\partial q_i}\left(h_{i+1}h_{i+2}\overbrace{\hat{u}_i\cdot\hat{u}_i}^{1}\hat{u}_jG_{ij}\right)$$

Since $\hat{u}_i\cdot\hat{u}_i = 1$, we have

$$\nabla\cdot\overline{\overline{G}} = \frac{1}{h_1h_2h_3}\sum_{i=1}^{3}\sum_{j=1}^{3}\frac{\partial}{\partial q_i}\left(h_{i+1}h_{i+2}G_{ij}\hat{u}_j\right)$$

and separating the derivative of a product by $\partial\left[\hat{u}_i(q_1,q_2,q_3)f(q_1,q_2,q_3)\right]/\partial q_j = \hat{u}_i(\partial f/\partial q_j) + (\partial\hat{u}_i/\partial q_j)f$,

$$\nabla\cdot\overline{\overline{G}} = \frac{1}{h_1h_2h_3}\sum_{i=1}^{3}\sum_{j=1}^{3}\left[\hat{u}_j\frac{\partial\left(h_{i+1}h_{i+2}G_{ij}\right)}{\partial q_i} + \left(\frac{\partial\hat{u}_j}{\partial q_i}\right)\left(h_{i+1}h_{i+2}G_{ij}\right)\right] \qquad (4.7\text{-}9)$$

This, then, is the general form for the divergence of the dyadic $\overline{\overline{G}}$, which we will use in formulating the vector Laplacian. Equations (4.7-8) and (4.7-9) are specialized for the case of Cartesian coordinates in Appendix B, Eqs. (B.1-5) and (B.1-4).

### 4.7.4(b) The vector Laplacian in GOCCs

Two features of the divergence of a dyadic must be considered. First, the inner-product rules illustrated in Eqs. (3.4-7)–(3.4-10) are applicable when applying the divergence; however, care must be taken to account for variations in both magnitude as well as direction of coordinate variables. Since the gradient of our vector field $\overline{A}(q_1, q_2, q_3)$ is given by the two double summations of Eq. (4.3-20), after some collection of common factors we have

$$\nabla\overline{A} = \sum_{i=1}^{3}\frac{\hat{u}_i}{h_i}\left[\sum_{j=1}^{3}\left(\hat{u}_j\frac{\partial A_j}{\partial q_i} + A_j\frac{\partial \hat{u}_j}{\partial q_i}\right)\right] \tag{4.7-10}$$

Substituting Eq. (4.7-10) into Eq. (4.7-7), we have

$$\nabla\cdot\nabla\overline{A} = \frac{1}{h_1h_2h_3}\sum_{i=1}^{3}\frac{\partial}{\partial q_i}\left\{h_{i+1}h_{i+2}(\hat{u}_i\cdot\hat{u}_i)\frac{1}{h_i}\left[\sum_{j=1}^{3}\left(\hat{u}_j\frac{\partial A_j}{\partial q_i} + A_j\frac{\partial \hat{u}_j}{\partial q_i}\right)\right]\right\}$$

and taking the inner product $\hat{u}_i\cdot\hat{u}_i = 1$ we have the vectorial resultant of the divergence of our dyadic gradient of the vector field $\overline{A}(q_1, q_2, q_3)$:

$$\nabla\cdot\nabla\overline{A} = \frac{1}{h_1h_2h_3}\sum_{i=1}^{3}\frac{\partial}{\partial q_i}\left\{\frac{h_{i+1}h_{i+2}}{h_i}\left[\sum_{j=1}^{3}\left(\hat{u}_j\frac{\partial A_j}{\partial q_i} + A_j\frac{\partial \hat{u}_j}{\partial q_i}\right)\right]\right\} \equiv \nabla^2\overline{A} \tag{4.7-11}$$

This is an expression for the vector Laplacian in GOCCs.

Because of the double summation above and the two terms of one of the summations, there are 18 terms that will need to be collected to determine the three vector components of the resultant. The second term inside of the parentheses represents the nine coordinate derivatives of unit vectors. Whereas these nine derivatives can be expanded with the use of Eqs. (1.3-19) and (1.3-20), we will not do so here because it unnecessarily complicates the analysis. Instead, we will illustrate how Eq. (4.7-11) is used in a specific coordinate system, namely cylindrical coordinates.

### 4.7.4(c) The vector Laplacian in cylindrical coordinates

As was pointed out in Section 4.3.2, there are nine unit vector derivatives. In cylindrical coordinates, seven of these nine are zero. The only nonzero derivatives are from Eq. (4.3-21) and Eq. (4.3-22), namely $\partial\hat{u}_r/\partial\phi = \hat{u}_\phi$ and $\partial\hat{u}_\phi/\partial\phi = -\hat{u}_r$. Substituting $r, \phi, z$ for $q_1, q_2, q_3$ and $1, r, 1$ for $h_1, h_2, h_3$ we have

$$\nabla\cdot\Big|_{cyl} = \frac{1}{r}\frac{\partial}{\partial r}\left(r\hat{u}_r\cdot\quad\right) + \frac{1}{r}\frac{\partial}{\partial\phi}\left(\hat{u}_\phi\cdot\quad\right) + \frac{\partial}{\partial z}\left(\hat{u}_z\cdot\quad\right) \tag{4.7-12}$$

The $1/(h_1h_2h_3) = 1/r$ factor in Eq. (4.7-7) appears in all three terms: however, the $h_1h_2 = r$ factor in the argument of the third differential cancels since the coordinate $r$ is orthogonal to $z$. We next rearrange Eq. (4.3-23), the dyadic gradient of the vector $\overline{A}(r,\phi,z)$ in cylindrical coordinates, for insertion into Eq. (4.7-10) by carefully orienting the nine unit dyads as follows:

$$\begin{aligned}\nabla\overline{A}\Big|_{cyl} &= +\hat{u}_r\left[\hat{u}_r\frac{\partial A_r}{\partial r} + \hat{u}_\phi\frac{\partial A_\phi}{\partial r} + \hat{u}_z\frac{\partial A_z}{\partial r}\right] \\ &\quad + \frac{\hat{u}_\phi}{r}\left[\hat{u}_r\left(\frac{\partial A_r}{\partial\phi} - A_\phi\right) + \hat{u}_\phi\left(\frac{\partial A_\phi}{\partial\phi} + A_r\right) + \hat{u}_z\frac{\partial A_z}{\partial\phi}\right] \\ &\quad + \hat{u}_z\left[\hat{u}_r\frac{\partial A_r}{\partial z} + \hat{u}_\phi\frac{\partial A_\phi}{\partial z} + \hat{u}_z\frac{\partial A_z}{\partial z}\right]\end{aligned} \tag{4.7-13}$$

Equation (4.7-13) is now ready for inclusion into Eq. (4.7-12). Notice that all we need to do is to insert the first, second and third terms of Eq. (4.7-13) into the first, second and third terms of Eq. (4.7-12), respectively, because the resulting unit vector dot products otherwise go to zero. This process results in

$$\begin{aligned}\nabla\cdot\nabla\overline{A}\Big|_{cyl} &= \frac{1}{r}\frac{\partial}{\partial r}\left[r\hat{u}_r\cdot\hat{u}_r\left(\hat{u}_r\frac{\partial A_r}{\partial r} + \hat{u}_\phi\frac{\partial A_\phi}{\partial r} + \hat{u}_z\frac{\partial A_z}{\partial r}\right)\right] \\ &\quad + \frac{1}{r}\frac{\partial}{\partial\phi}\left\{\hat{u}_\phi\cdot\hat{u}_\phi\frac{1}{r}\left[\hat{u}_r\left(\frac{\partial A_r}{\partial\phi} - A_\phi\right) + \hat{u}_\phi\left(\frac{\partial A_\phi}{\partial\phi} + A_r\right) + \hat{u}_z\frac{\partial A_z}{\partial\phi}\right]\right\} \\ &\quad + \frac{\partial}{\partial z}\left[\hat{u}_z\cdot\hat{u}_z\left(\hat{u}_r\frac{\partial A_r}{\partial z} + \hat{u}_\phi\frac{\partial A_\phi}{\partial z} + \hat{u}_z\frac{\partial A_z}{\partial z}\right)\right]\end{aligned}$$

Before rushing into replacing the $\hat{u}_i\cdot\hat{u}_i$ factors with unity, the student new to the tensor world should take note that in performing the $\hat{u}_i\cdot\hat{u}_i = 1$ process the rank is reduced by two. This means that the dot in the divergence operator has the same effect in the determination of the rank of the resultant as it has in the dot-product operation between any two tensors. In this case, we have a vector operator acting on a dyadic through a dot-product type of an operation, called the divergence, having the same effect of a vector dotted with a dyadic as far as the resultant rank is concerned. Before taking the unit vector dot products, one should realize that there were 27 unit triads $\hat{u}_i \cdot \hat{u}_j \hat{u}_k$. Eighteen of these go to zero when $i \neq j$ and the remaining nine are left as noted above. Now letting $\hat{u}_i \cdot \hat{u}_i = 1$, we have nine remaining terms, which after performing the appropriate unit vector derivatives, can be compiled into three components making up the resultant vector form of our vector Laplacian.

Since the unit vectors are invariant to differentiation with respect to $r$ and $z$, they may be taken out of the differential arguments of the first and third major terms of Eq. (4.7-12). This is also true of the $\hat{u}_z$ component of the second major term. Therefore, our vector Laplacian takes the intermediate form

$$\begin{aligned}\nabla \cdot \nabla \overline{A}\Big|_{cyl} &= \hat{u}_r \frac{1}{r}\frac{\partial}{\partial r}\left(r\frac{\partial A_r}{\partial r}\right) + \hat{u}_\phi \frac{1}{r}\frac{\partial}{\partial r}\left(r\frac{\partial A_\phi}{\partial r}\right) + \hat{u}_z \frac{1}{r}\frac{\partial}{\partial r}\left(r\frac{\partial A_z}{\partial r}\right)\\ &+ \frac{1}{r^2}\frac{\partial}{\partial \phi}\left[\hat{u}_r\left(\frac{\partial A_r}{\partial \phi} - A_\phi\right) + \hat{u}_\phi\left(\frac{\partial A_\phi}{\partial \phi} + A_r\right)\right] + \hat{u}_z \frac{1}{r^2}\frac{\partial^2 A_z}{\partial \phi^2}\\ &+ \hat{u}_r \frac{\partial^2 A_r}{\partial z^2} + \hat{u}_\phi \frac{\partial^2 A_\phi}{\partial z^2} + \hat{u}_z \frac{\partial^2 A_z}{\partial z^2}\end{aligned}$$

where the $\phi$ derivatives of $\hat{u}_r$ and $\hat{u}_\phi$ in the square brackets above need to be treated carefully. Substituting $\partial \hat{u}_r / \partial \phi = \hat{u}_\phi$ and $\partial \hat{u}_\phi / \partial \phi = -\hat{u}_r$, the $\phi$ derivative becomes

$$\begin{aligned}&\frac{\partial}{\partial \phi}\left[\hat{u}_r\left(\frac{\partial A_r}{\partial \phi} - A_\phi\right) + \hat{u}_\phi\left(\frac{\partial A_\phi}{\partial \phi} + A_r\right)\right] =\\ &\hat{u}_r\left(\frac{\partial^2 A_r}{\partial \phi^2} - \frac{\partial A_\phi}{\partial \phi}\right) + \underbrace{\frac{\partial \hat{u}_r}{\partial \phi}}_{\hat{u}_\phi}\left(\frac{\partial A_r}{\partial \phi} - A_\phi\right) + \hat{u}_\phi\left(\frac{\partial^2 A_\phi}{\partial \phi^2} + \frac{\partial A_r}{\partial \phi}\right) + \underbrace{\frac{\partial \hat{u}_\phi}{\partial \phi}}_{-\hat{u}_r}\left(\frac{\partial A_\phi}{\partial \phi} + A_r\right)\end{aligned}$$

Collecting vector components and applying Eq. (4.7-6) to the scalar Laplacian of the scalar component of the operand vector $\overline{A} = \hat{u}_1 A_1 + \hat{u}_2 A_2 + \hat{u}_3 A_3$, we obtain the vector Laplacian in cylindrical coordinates:

$$\nabla^2 \overline{A}\Big|_{cyl} = \hat{u}_r\left(\nabla^2 A_r - \frac{2}{r^2}\frac{\partial A_\phi}{\partial \phi} - \frac{A_r}{r^2}\right) + \hat{u}_\phi\left(\nabla^2 A_\phi + \frac{2}{r^2}\frac{\partial A_r}{\partial \phi} - \frac{A_\phi}{r^2}\right) + \hat{u}_z \nabla^2 A_z \tag{4.7-14}$$

consistent with Ramo, Whinnery, and Van Duzer (inside the front cover),[14] for example. Notice that this expansion is not simply the vector sum of the individually formed Laplacians of the scalar components of the vector $\overline{A}$ in cylindrical coordinates. Since the usual approach in presenting the vector Laplacian to undergraduate students is to use Cartesian coordinates, where $\nabla^2\overline{A} = \hat{u}_x\nabla^2 A_x + \hat{u}_y\nabla^2 A_y + \hat{u}_z\nabla^2 A_z$, the extra terms in the radial and azimuthal components of Eq. (4.7-14) that stem from the coordinate derivatives of the unit vectors would not at all be evident. This is a common mistake by students when the expansion is carried out in other than Cartesian coordinates.

### 4.7.5 The curl of the curl of a vector field and the Lagrange identity

There are two additional second-order vector differential operators, both of which have vector operands, that are commonly used in upper-division courses. These are the curl of the curl (**CCv** $= \nabla\times\nabla\times\overline{A}$) and the gradient of the divergence (**GDv** $= \nabla\nabla\cdot\overline{A}$). **CCv** is discussed in this section, while **GDv** is discussed in the next. These two are important because of the *Lagrange vector identity*:

$$\nabla^2\overline{A} = \nabla(\nabla\cdot\overline{A}) - \nabla\times\nabla\times\overline{A} \tag{4.7-15}$$

The left-hand side of Eq. (4.7-15)—the vector Laplacian—is essential in the formulation of vector wave equations used in junior-level electromagnetics, quantum physics, and other similar courses. This identity is presented to the undergraduate student in lieu of having to deal with the dyadic gradient of a vector as well as the divergence of the resulting dyadic, which, as we saw from the previous section, were used in determining the vector Laplacian. Since none of the operations on the right side of Eq. (4.7-15) require any consideration of tensors of higher rank than unity, namely a scalar and three vector operands, the vector Laplacian can be determined without the need for dyadics by using this identity.

The pedagogical problem with this approach, however, is that (4.7-15) cannot be proven, even in Cartesian coordinates (which would be quite adequate), without dealing with the dyadic operations discussed above. Therefore, we will develop the right side of Eq. (4.7-15) and show that it equals

Eq. (4.7-11). In this way, a rigorous proof of Eq. (4.7-15) is provided. Although mathematical identities may be proven in any coordinate system without loss of generality, we conduct this proof in GOCCs* since our objective here is primarily to develop first- and second-order expansions of our del operators for conversion to any other orthogonal system appropriate to the natural geometry of the problem. However, before expanding the curl of the curl needed for the last term of Eq. (4.7-15) we first provide a physical description.

### 4.7.5(a) A physical description of the curl of the curl

In describing the curl operator (in Section 4.5.2) as a measure of the *circulation density* or *vorticity* of a vector field, we expand on that description in giving a physical description of the curl of the curl as follows:

> *The curl of the curl of a vector field is the circulation density of the vorticity of that field, which can be thought of as the rotational spatial change of vorticity in the cross-product direction.*

What is meant by the "cross-product direction" is that its direction is generally at a large acute angle (nearly orthogonal, but not necessarily at right angles) to the vorticity, which in turn may either be in the direction of the original vector field or "nearly orthogonal" to both.

Let us illustrate this concept by two simple hypothetical examples, both dealing with friction-free circulating liquids in an upright cylindrical tub. The first has a uniform circulating density field, that is, it has no variation in its circulation. Thus, it has a constant curl and therefore a zero curl curl. In the second example, the circulation density in the center is greater than on the outside, as if the liquid were draining from a hole in the center of the tub, forming a cyclonic-type of a vortex hole in the flow field. We will oversimplify the vector fields involved in this example to illustrate the point of a nonzero curl curl.

### *Example 1: Vorticity in a uniform angular velocity field*

Suppose the velocity of the liquid is represented by the vector field

$$\overline{v} = \hat{u}_\phi ar \tag{4.7-16}$$

---

* Sections 4.7.5(b) and 4.7.7 outline the process of this proof; however, the rigorous proof is left to Appendix C.

where $a$ is a constant and $r$ and $\phi$ are the radial and azimuthal cylindrical coordinates, with the $z$-coordinate falling on the axis of the cylinder in the right-hand sense. This says that the entire body of liquid rotates in unison (as if the liquid were stationary and the tub were rotating). The curl of $\overline{v}$ is

$$\nabla \times \overline{v} = \hat{u}_z \frac{1}{r}\frac{\partial (r v_\phi)}{\partial r} = \hat{u}_z \frac{1}{r}\frac{\partial (a r^2)}{\partial r} = \hat{u}_z 2a \qquad (4.7\text{-}17)$$

which is a vector with constant direction ($\hat{u}_z$) and magnitude ($2a$) everywhere in the region. That is, if a curl meter [described in the third paragraph of Section (4.5.2)] were placed in the rotating fluid with its axis parallel to the $z$ axis, it will rotate counterclockwise with the same rotational velocity $\overline{\omega}$ at all positions in the region. The direction of the vector field $\overline{\omega}$ is, of course, that of the axis of the curl meter in the right-hand sense, namely, $\hat{u}_z$.

The vector $\overline{\omega}$ is called the *vorticity vector* and is defined as half of the curl of the velocity field:

$$\overline{\omega} = \tfrac{1}{2}\nabla \times \overline{v} = \hat{u}_z a \qquad (4.7\text{-}18)$$

which says that the vorticity is uniform everywhere in the region, as anticipated by our *gedanken* experiment of the previous paragraph. Next, we take the curl of $\overline{\omega}$ and find that it is zero because it has no variation. Thus,

$$\nabla \times \nabla \times \overline{v} = 0 \qquad (4.7\text{-}19)$$

which serves as our example of nonrotational vorticity, in other words, nonvarying vorticity.

### *Example 2: Vorticity in a nonuniform angular velocity field*

Let us next express the velocity field for our rotating liquid by

$$\overline{v} = \hat{u}_\phi 2a \qquad (4.7\text{-}20)$$

Here the azimuthal velocity is the same no matter the radius. This means that angular velocity must vary as $1/r$, i.e., at half of the radius, the angular velocity doubles in order for the linear velocity $v_\phi$ to remain constant. (This simplistic example ignores the centripetal behavior of the mass of the liquid as the radius goes to zero, but is used to illustrate the basic nature of finite curl curl). The curl of Eq. (4.7-20) is

$$\nabla\times\overline{v} = \hat{u}_z \frac{1}{r}\frac{\partial(rv_\phi)}{\partial r} = \hat{u}_z \frac{2a}{r} \tag{4.7-21}$$

which is everywhere directed axially ($\hat{u}_z$) but varies with radius hyperbolically, namely as $2a/r$. The vorticity, then, is $\overline{\varpi} = \hat{u}_z\, a/r$, which increases in magnitude toward the center. Taking the curl again, we have a description of the circulation density of the vortex field

$$\nabla\times\nabla\times\overline{v} = -\hat{u}_\phi \frac{\partial\left(\frac{2a}{r}\right)}{\partial r} = \hat{u}_\phi \frac{2a}{r^2} \tag{4.7-22}$$

which clearly is not zero. Thus, the curl of the vorticity is $\nabla\times\overline{\varpi} = \hat{u}_\phi\, a/r^2$, which increases quadratically with decreasing radius, giving the semblance (but not the exact formulation) of a cyclonic-type of a vortex hole in the flow field.

### 4.7.5(b) The curl of the curl in GOCCs

By applying Eq. (4.5-12a) twice we have

$$\begin{aligned}
\nabla\times\nabla\times\overline{A} = \hat{u}_1 \frac{1}{h_2h_3}\Bigg( & \frac{\partial}{\partial q_2}\left\{\frac{h_3}{h_1h_2}\left[\frac{\partial(h_2A_2)}{\partial q_1} - \frac{\partial(h_1A_1)}{\partial q_2}\right]\right\} \\
& - \frac{\partial}{\partial q_3}\left\{\frac{h_2}{h_3h_1}\left[\frac{\partial(h_1A_1)}{\partial q_3} - \frac{\partial(h_3A_3)}{\partial q_1}\right]\right\}\Bigg) \\
+ \hat{u}_2 \frac{1}{h_3h_1}\Bigg( & \frac{\partial}{\partial q_3}\left\{\frac{h_1}{h_2h_3}\left[\frac{\partial(h_3A_3)}{\partial q_2} - \frac{\partial(h_2A_2)}{\partial q_3}\right]\right\} \\
& - \frac{\partial}{\partial q_1}\left\{\frac{h_3}{h_1h_2}\left[\frac{\partial(h_2A_2)}{\partial q_1} - \frac{\partial(h_1A_1)}{\partial q_2}\right]\right\}\Bigg) \\
+ \hat{u}_3 \frac{1}{h_1h_2}\Bigg( & \frac{\partial}{\partial q_1}\left\{\frac{h_2}{h_3h_1}\left[\frac{\partial(h_1A_1)}{\partial q_3} - \frac{\partial(h_3A_3)}{\partial q_1}\right]\right\} \\
& - \frac{\partial}{\partial q_2}\left\{\frac{h_1}{h_2h_3}\left[\frac{\partial(h_3A_3)}{\partial q_2} - \frac{\partial(h_2A_2)}{\partial q_3}\right]\right\}\Bigg)
\end{aligned} \tag{4.7-23a}$$

or alternatively, by applying Eq. (4.5-12b) twice, the curl of the curl becomes

$$\nabla\times\nabla\times\bar{A}=\begin{vmatrix} \dfrac{\hat{u}_1}{h_2h_3} & \dfrac{\hat{u}_2}{h_3h_1} & \dfrac{\hat{u}_3}{h_1h_2} \\ \dfrac{\partial}{\partial q_1} & \dfrac{\partial}{\partial q_2} & \dfrac{\partial}{\partial q_3} \\ \dfrac{h_1}{h_2h_3}\left[\dfrac{\partial}{\partial q_2}(h_3A_3)-\dfrac{\partial}{\partial q_3}(h_2A_2)\right] & \dfrac{h_2}{h_3h_1}\left[\dfrac{\partial}{\partial q_3}(h_1A_1)-\dfrac{\partial}{\partial q_1}(h_3A_3)\right] & \dfrac{h_3}{h_1h_2}\left[\dfrac{\partial}{\partial q_1}(h_2A_2)-\dfrac{\partial}{\partial q_2}(h_1A_1)\right] \end{vmatrix} \quad (4.7\text{-}23b)$$

where the expansion of Eq. (4.7-23b) is readily shown to become Eq. (4.7-23a). Equation (4.7-23b) is consistent with Stratton [page 50].[17]

### 4.7.6 The gradient of the divergence of a vector field

Recall that the impetus for exploring the curl of the curl in the previous section (and the gradient of the divergence in this section) was in part due to the Lagrange identity, Eq. (4.7-15), $\nabla^2\bar{A}=\nabla(\nabla\cdot\bar{A})-\nabla\times\nabla\times\bar{A}$ that is, (**DGv** = **GDv-CCv**). The left-hand side was developed in detail in Section 4.7.4. This required that one determine the gradient of a vector (**Gv**) and take the divergence of the resulting dyadic. The right-hand side of this identity provides the vector Laplacian entirely by the use of vector differential operators with only scalar and vector operands.

However, a usual approach (especially at the undergraduate level) is to use this identity without proof. Even if the proof is conducted in Cartesian coordinates, which would be quite adequate, the dyadic resultant from the **Gv** operation and the divergence of the resulting dyadic are still needed. In the past, the undergraduate has not been expected to deal with such matters. Thus, as a first step toward the proof and in the interest of completeness, in Section 4.7.5 we developed the second term on the right-hand side of this identity, **CCv**. Here we will explore the first term, **GDv**.

### 4.7.6(a) A physical description of the gradient of the divergence

First, the divergence of a vector field is equal to the net outward flux from infinitesimal closed surfaces at every point in space where the divergence is desired. Since that value is the volume source distribution density at each said point,

> *then the gradient of the divergence of a vector field is another vector field oriented in the direction in which the volume source distribution density increases most rapidly. Its magnitude is the derivative of that volume density distribution in the direction of its maximal increase.*

The existence of the gradient of the divergence is dependent upon the existence of such a directional derivative. If there are no sources at the point in space where the grad-div is being calculated, then **GDv**=0.

### 4.7.6(b) The gradient of the divergence in GOCCs

By replacing the scalar $V$ in Eq. (4.3-19) with the scalar divergence of Eq. (4.4-22), we have

$$\nabla\nabla\cdot\bar{A} = \sum_{i=1}^{3}\hat{u}_i\frac{1}{h_i}\frac{\partial}{\partial q_i}\left[\sum_{j=1}^{3}\frac{1}{h_1h_2h_3}\frac{\partial}{\partial q_j}\left(\frac{h_1h_2h_3}{h_j}A_j\right)\right] \tag{4.7-24}$$

## 4.7.7 The gradient of the divergence minus the curl of the curl—the vector Laplacian

Subtracting Eq. (4.7-23) from Eq. (4.7-24) leads to Eq. (4.7-11) after some development:

$$\nabla\nabla\cdot\bar{A}-\nabla\times\nabla\times\bar{A}=\frac{1}{h_1h_2h_3}\sum_{i=1}^{3}\frac{\partial}{\partial q_i}\left\{\frac{h_{i+1}h_{i+2}}{h_i}\left[\sum_{j=1}^{3}\left(\hat{u}_j\frac{\partial A_j}{\partial q_i}+A_j\frac{\partial \hat{u}_j}{\partial q_i}\right)\right]\right\}$$
$$=\nabla^2\bar{A} \tag{4.7-25}$$

This is demonstrated in Appendix C.

## References

1. William E. Boyce and Richard C. DiPrima, *Elementary Differential Equations*, 5th ed., Wiley, New York (1992).
2. Peter V. O'Neill, *Advanced Engineering Mathematics*, 3rd ed., Wadsworth, Belmont, CA (1983).
3. David A. Sanchez, *Ordinary Differential Equations—A Brief Eclectic Tour*, The Mathematical Association of America, Washington, DC (2002).
4. George B. Thomas, Jr. and Ross L. Finney, *Calculus and Analytic Geometry*, 8th ed., Addison-Wesley, Reading, MA (1992).
5. James Stewart, *Calculus*, 2nd ed., Brooks/Cole, Monterey, CA (1991).
6. Gilles Aubert and Pierre Kornprobst, *Mathematical Problems in Image Processing—Partial Differential Equations and the Calculus of Variations*, Springer-Verlag, Berlin (2002).
7. Cargill Gilston Knott, in *Life and Scientific Work of Peter Guthrie Tait*, Cambridge University Press (1911).
8. W. Michael Lai, David Rubin, and Erhard Kremple, *Introduction to Continuum Mechanics*, Pergamon Press, New York (1974).
9. A. I. Lur'e and D. B. McVean, *Three Dimensional Problems of Elasticity*, Wiley Interscience, Hoboken, NJ (1964).
10. Robert W. Boyd, *Nonlinear Optics*, Academic Press, London (1992).
11. Govind P. Agrawal, *Nonlinear Fiber Optics*, 3rd ed., Academic Press, London (2001).
12. Vladislav Bevc, "Vector differential operations derived from physical definitions", *Am. J. Phys.* **34**(6), 507-510 (June 1966).
13. Robert H. Romer, Editor, "Editorial: Memorable papers from the American Journal of Physics, 1933-1990", *Am. J. Phys.* **59**(3), 201-207 (Mar. 1991).
14. Simon Ramo, John R. Whinnery, and Theodore Van Duzer, *Fields and Waves in Communication Electronics*, 3rd ed., Wiley, New York (1994).
15. Philip M. Morse and Herman Feshbach, *Methods of Theoretical Physics*, McGraw-Hill, New York (1953).
16. Walter E. Rogers, *Introduction to Electric Fields*, McGraw-Hill, New York (1954).

17. Julius Adams Stratton, *Electromagnetic Theory*, McGraw-Hill, New York (1941).

## Chapter 5

# Vector Calculus Integral Forms

There is an intimate relationship between differential and integral forms in vector calculus (and tensor calculus as well). For example, Maxwell's curl equations for time-varying electric and magnetic field intensities, which are vector *differential* operators, convert to circulations of these time-varying fields, which are *integral* forms that describe the electromotive and magnetomotive force (volts and amps), respectively. Further, Maxwell's divergence equations for the electric and magnetic flux densities (*differential* forms) convert to closed-surface *integral* forms. These conversion relationships can be developed from a series of theorems from the mathematics of George Green (1828) called *Green's identities*.

Other mathematicians of the 1800s contributed various forms of identities—such as Gauss' and Stokes' theorems, discussed in Sections 5.3 and 5.4, respectively—that significantly add to the tools for converting between differential and integral forms. Since Gauss' work preceded Green's, it would be accurate to describe the relevant Green's forms as generalizations of Gauss'; and since Stokes' theorem followed Green's, one could take the position that Stokes' theorem is a special case of one of Green's identities.

Green's mathematics also included the *Green's function*, which provides an effective method for determining solutions to inhomogeneous differential equations. This process will be covered in Section 5.5; for now, it is sufficient to say that this tool further provides evidence of this differential-integral relationship.

Before probing into the powerful mathematics of these forms and theorems, we first elaborate on line and surface integrals for two reasons. First, all of the aforementioned theorems involve line or surface integrals or both. Secondly, this elaboration will provide comprehensiveness so that the breadth of physical applications may be described. In Section 2.4, line and surface integrals were introduced as examples of integrands made up of the vector dot product of a vector field with the vector line and surface differentials $\overline{d\ell}$ and $\overline{da}$ [see Eqs. (2.4-20)–(2.4-23)]. In Sections 5.1 and 5.2, line and surface integrals, respectively, are covered more generally.

Volume integrals are, of course, involved in the 3D forms of Green's identities, including Gauss' theorem, a special case of one of the Green's identities. However, since differential volume is inherently scalar (see Section 1.2.3), volume integrals do not modify the rank. By that, we mean that the resultant quantity after performing the integration has the same rank as the integrand because the differential element of the integration $dv$ is a scalar. For this reason, we choose not to expand on volume integrals as we have in the next two sections for line and surface integrals, which have vector differentials $\overline{d\ell}$ and $\overline{da}$ and can indeed change the resultant rank.

## 5.1 Line Integrals of Vector (and Other Tensor) Fields

Whereas $\int \overline{F} \cdot \overline{d\ell}$ [Eq. (2.4-20)] is the specific line integral of the tangential component of a vector field along a prespecified line in 3D space, many other line integrals exist. These are shown in the subsection below for thoroughness. Examples are then given for the specific form of Eq. (2.4-20).

### 5.1.1 Line integrals of scalar, vector, and tensor fields with dot-, cross-, and direct-product integrands

A general definition of a line integral:

*An integral of a field quantity taken over a vector differential length $\overline{d\ell}$ that is everywhere tangent to a general line $L$ in space is a line integral.*

Notice that the "tangent" in this definition refers to the line segment $\overline{d\ell}$ and not to the field, the latter of which may have general direction and be of general rank. There are several line integral forms of scalar-, vector- and tensor-field integrands, each with three types of product operations. Several of these line-integral forms are shown below in the order of their resultant rank, noted to the right of each form.

$$\int_L \overline{F}(\overline{r}) \cdot \overline{d\ell} \qquad \text{scalar, } n_R = 0 \qquad (5.1\text{-}1)$$

$$\int_L f(\overline{r}) \overline{d\ell} \qquad \text{vector, } n_R = 1 \qquad (5.1\text{-}2)$$

$$\int_L \overline{F}(\overline{r}) \times \overline{d\ell} \qquad \text{vector, } n_R = 1 \qquad (5.1\text{-}3)$$

where the fields $\overline{F}(\overline{r})$ and $f(\overline{r})$ are written here in generalized $\overline{r}$-space notation as described is Section 1.1.4. We will next discuss these first three line-integral forms and then continue with the list ending in three line-integral forms involving tensor integrands in general.

Expression (5.1-1) is of the form of Eq. (2.4-20) cited earlier. This integral is the sum of the tangential components of $\overline{F}$ along $L$ through the dot-product operation. This particular operation yields a scalar field commonly referred to as the potential field. This form is a major part of this section because of its frequency of use in electromagnetics, photonics, and materials science. As such, this form will be covered in greater depth with examples in Section 5.1.2.

Line-integral forms (5.1-2) and (5.1-3) are also commonly used in fields and photonics. These are the vector sum of the direct product of a scalar field $f(\overline{r})$ with each vector differential element $\overline{d\ell}$ along the path $L$—form (5.1-2), and vector sum of perpendicular components of a vector field $\overline{F}$ along $L$ through the cross-product operation—form (5.1-3). Both of these forms result in vector fields. Section 5.1.3 provides examples of the utility of these forms for the case of the magnetic vector potential and the magnetic field intensity $\overline{H}$, respectively, resulting from a filamentary electric current source.

Before listing several other line-integral forms, it should be pointed out that integrals (5.1-1) through (5.1-3) are presented by McQuistan[1] with several examples of their use in Cartesian coordinates. We will use a variation of one of these examples in describing the properties of expression (5.1-1). (See Section 5.1.2.) A study of McQuistan's examples of forms (5.1-2) and (5.1-3) are highly recommended because they provide further examples with excellent physical interpretations.

The remaining line-integral forms are listed here for completeness and for citing further examples.

Whereas the first example [expression (5.1-1)] has a scalar-field resultant, the next three, expressions (5.1-2) through (5.1-4), have vector-field resultants. The first two of these vector-resultant forms, namely expressions (5.1-2) and (5.1-3), have already been listed above with references cited to examples given in the following two subsections. The third is

$$\int_L \overline{\overline{A}}(\overline{r})\cdot\overline{d\ell} \qquad \text{vector, } n_R = 1 \qquad (5.1\text{-}4)$$

where nine inner-product operations of each of the components of the dyadic field $\overline{\overline{A}}(\overline{r})$, *including their unit dyads*, are taken with the vector components of $\overline{d\ell}$ before the nine integral operations are made. This process in the integrand is identical to the process described in Section 3.4.1(a)—the dyadic dot product with a vector—resulting in a vector field. Recall that this process illustrated the nature of the "inner (dot) product" in that the "application of the inner product eliminates two of the three unit vectors thereby reducing the sum of the ranks of the two quantities involved by two." See rule #3 in Section 3.4.1(a). An application of Eq. (5.1-4) is presented in Appendix D following Eq. (D.1-5).

As with expression (5.1-1), the "dot $\overline{d\ell}$" in expression (5.1-4) also implies that the tangential components of the dyadic $\overline{\overline{A}}$ are taken along the path $L$. However, the components of a dyadic are dual directional. So naturally, one new to tensors might ask what is meant by "tangential component" in the context of dyadics (or tensors in general). Recall that the unit dyads $\hat{\hat{u}}_{ij}$ were introduced in Eq. (1.1-6b), but before they could be viewed explicitly as an inner-product operation [Eq. (3.4-1)], they had to be expanded as $\hat{u}_i\hat{u}_j$ per Eq. (3.3-4). The order of this expansion was important because the latter unit vector, namely $\hat{u}_j$ (and not $\hat{u}_i$), was dotted with the vector components in the process of obtaining the vector result [Eq. (3.4-3)]. Therefore, when we say that the tangential components of $\overline{\overline{A}}$ are taken along the path $L$, we are referring to the latter unit vector in the nine unit dyads. The resultant is a vector that in general is not aligned with $\overline{d\ell}$.

Further, for tensors of higher rank, such as in form (5.1-7), where we are taking the inner product of a triad $\overline{\overline{\overline{T}}}$ with the vector differential length segment $\overline{d\ell}$, in each of the 27 components the unit triad $\hat{\hat{\hat{u}}}_{ijk}$ must be expanded as $\hat{\hat{u}}_{ij}\hat{u}_k$ or $\hat{u}_i\hat{u}_j\hat{u}_k$. In this way, the last unit vector, namely $\hat{u}_k$, is ready for the inner-product operation with the vector $\overline{d\ell}$. Incidentally, the opposite is the case if the vector and tensor are interchanged. That is, in $\overline{A}\cdot\overline{\overline{\overline{T}}}$ the first unit vector in $\hat{\hat{\hat{u}}}_{ijk}$ must be isolated, namely $\hat{u}_i\hat{\hat{u}}_{jk}$, in preparation for the inner-product operation of $\overline{A}$ with $\overline{\overline{\overline{T}}}$. This point was made explicitly in Section 3.4.1(b).

Following the three vector-resultant line-integral forms, the next three forms result in dyadic fields, expressions (5.1-5) through (5.1-7). The first of these is the integral of the "direct" product of the vector field $\overline{F}(\overline{r})$ with the vector differential $\overline{d\ell}$ over the path $L$, which of course, is a dyadic field.

$$\int_L \overline{F}(\overline{r})\overline{d\ell} \equiv \int_L \overline{F}(\overline{r}) \otimes \overline{d\ell} \qquad \text{dyadic } n_R = 2 \qquad (5.1\text{-}5)$$

(See the first footnote on page 4-3 for an explanation of the equivalence between "direct product" and "tensor product.") The next two line integrals deal with integrands that also result in dyadics following the rules outlined in the fifth paragraph of Section 4.1. These are the dyadic cross product (external product) and the triadic dot product (inner product) each with our vector differential $\overline{d\ell}$ as follows

$$\int_L \overline{\overline{A}}(\overline{r}) \times \overline{d\ell} \qquad \text{dyadic } n_R = 2 \qquad (5.1\text{-}6)$$

$$\int_L \overline{\overline{\overline{T}}}(\overline{r}) \cdot \overline{d\ell} \qquad \text{dyadic } n_R = 2 \qquad (5.1\text{-}7)$$

This process continues, but in general we can summarize all line-integral forms by taking the dot-, cross-, and (tensor-) direct-product operations of general rank tensors as

$$\int_L [{}_{n_R}T_{\mathbf{O}}(\overline{r})] \cdot \overline{d\ell} \quad \Rightarrow \quad {}_{n_R-1}T_{\mathbf{R}} \qquad (5.1\text{-}8)$$

$$\int_L [{}_{n_R}T_{\mathbf{O}}(\overline{r})] \times \overline{d\ell} \quad \Rightarrow \quad {}_{n_R}T_{\mathbf{R}} \qquad (5.1\text{-}9)$$

$$\int_L [{}_{n_R}T_{\mathbf{O}}(\overline{r})] \otimes \overline{d\ell} \quad \Rightarrow \quad {}_{n_R+1}T_{\mathbf{R}} \qquad (5.1\text{-}10)$$

where the post-subscripts "**O**" and "**R**" simply distinguish the *operand* tensor from the three *resultant* tensors. Thus, $[{}_{n_R}T_{\mathbf{O}}]$ is an operand tensor of rank $n_R$ and $[{}_iT_{\mathbf{R}}]$ is a resultant tensor of rank $i = n_R - 1$, $n_R$, or $n_R + 1$ for dot-, cross-, or direct-product integral operators, respectively.

### 5.1.2 Examples of form (5.1-1): Line integral of the tangential component of $\overline{F}$ along path *L*

The open line integral [expression (5.1-1)] was touched upon in Eq. (2.4-20) as an example of the application of dot product between a vector field and a differential line element used as an integrand. This integral takes the form $\int \overline{F} \cdot \overline{d\ell}$ and is commonly referred to as *the* line integral in common vector-calculus parlance. However, as one can see from expressions (5.1-3) and (5.1-5), there are other line integrals involving vector fields. Whereas expression (5.1-1) yields a scalar field, expression (5.1-3) yields another vector field and expression (5.1-5) yields a dyadic field. Thus, the typical use of the term "line integral" to mean expression (5.1-1) without mention of these other two vector-field line-integral operators is incomplete. It is mentioned here because it is in fact useful for particular applications that the scientist or engineer using this guide will encounter. It is also mentioned here for completeness.

In Eqs. (5.1-1), (5.1-3), and (5.1-5) $\overline{F}$ is a "force" field in a general sense. Generally speaking, a force field is a vector field that is causal. Its effect can be a scalar field as in Eq. (5.1-1), another vector field as in Eq. (5.1-3), or a dyadic field as in Eq. (5.1-5). In the case of Eq. (5.1-1), its scalar effect is that of a generalized potential. The examples below will clarify this point.

### 5.1.2(a) Examples in mechanics—force and work

In mechanics, including gravitational mechanics, $\overline{F}$ is a force field and is given the units of force, such as newtons in SI units. If $\overline{d\ell}$ is a differential line segment tangent to a predetermined line or path $L$ of action, then the differential work done by the force field is the scalar $dW = \overline{F} \cdot \overline{d\ell}$ (N·m or J). The total work between two points *a* and *b* on $L$ is

$$W_{ab} = \int_{a\,L}^{b} \overline{F}(\overline{r}) \cdot \overline{d\ell} \text{ joules} \tag{5.1-11}$$

where all differential line segments $\overline{d\ell}$ lie on $L$, which, in general, is an arbitrary continuous line in 3D space. By "continuous," we mean that the directional derivative is piecewise determinable. If there is a discontinuity in the directional derivative at a finite number of points $b_1, b_2, b_3, \cdots, b_N$ on $L$, between *a* and *b*, Eq. (5.1-11) must be broken into $N+1$ integrals as follows

$$W_{ab} = \int_{a\,L}^{b_1} \overline{F}(\overline{r}) \cdot \overline{d\ell} + \sum_{i=1}^{N-1} \int_{b_i\,L}^{b_{i+1}} \overline{F}(\overline{r}) \cdot \overline{d\ell} + \int_{b_N\,L}^{b} \overline{F}(\overline{r}) \cdot \overline{d\ell} \tag{5.1-12}$$

Equations (5.1-11) and (5.1-12) represent the work (energy) done *by* the force field $\overline{F}$ *on* an object in moving that object along the path $L$ from *a* to *b*. This is the decrease in potential energy and is referred to as the *potential energy difference*. By conservation of energy laws, this decrease in potential energy is transferred to the kinetic energy of the object less losses.

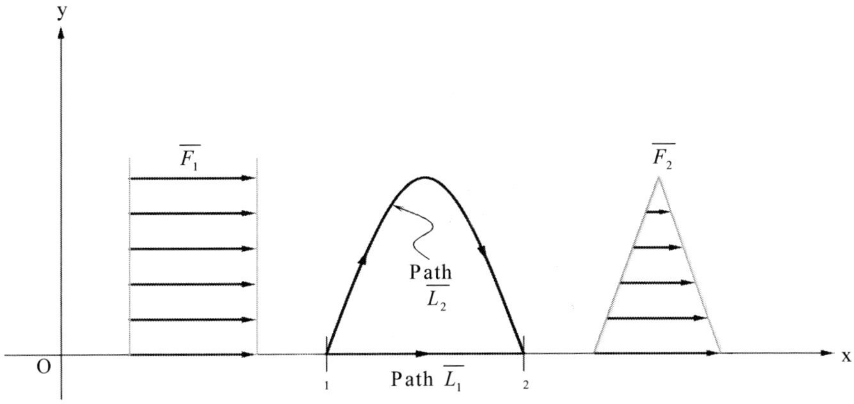


**Figure 5.1-1** Two integration paths, $L_1[1 \le x \le 2, y = 0, z = 0]$ and $L_2[1 \le x \le 2, y = 1-(3-2x)^2, z = 0]$ taken in a tangential-component-line integral from point *a*(1,0,0) to point *b*(2,0,0) in two force fields.

Two examples follow. The first example, although trivial, shows how Eq. (5.1-11) can be applied and illustrates a path-independent case. The second example is also simple, but illustrates path dependence. For both examples, let us pick two paths $L_1$ and $L_2$ each starting at $a(1,0,0)$ (meters) and ending at $b(2,0,0)$ (meters) as shown in Figure 5.1-1. Path $L_1$ lies on the $x$ axis and can be expressed as $L_1[1 \le x \le 2, y = 0, z = 0]$. Path $L_2$ is described as $L_2[1 \le x \le 2, y = 1-(3-2x)^2, z = 0]$, which takes a parabolic route in the $z = 0$ plane in going from $a$ to $b$.

## Example 1: Path-independent case

Suppose $\overline{F_1}$ is a uniform force field of one Newton in the $x$ direction, i.e., $\overline{F} = \hat{u}_x$, then $W_{ab}$ would be one N·m (or one J) by inspection. In this specific example, this result is independent of the path taken to get from $a(1,0,0)$ to $b(2,0,0)$, as will be shown next.

In general, $\overline{d\ell} = \hat{u}_x dx + \hat{u}_y dy$. Then, $\overline{F_1} \cdot \overline{d\ell} = \hat{u}_x \cdot \hat{u}_x dx + \hat{u}_x \cdot \hat{u}_y dy = dx$ and

$$W_{ab_{L1}} = W_{ab_{L2}} = \int_1^2 dx = 1 \text{ joule} \qquad (5.1\text{-}13)$$

in both cases. Thus, only the $x$ component of $\overline{d\ell}$ matters in this example. Further, the result is independent of the path taken.

*Example 2: Path-dependent case*

Suppose $\overline{F}_2$ varies with $y$ but retains its $x$ direction as $\overline{F}_2 = \hat{u}_x(1-y)$. At $y=0$, $\overline{F}_2 = \hat{u}_x$, which is the same as $\overline{F}_1$. Therefore, the work done on path $L_1$ is the same as in Example 1, namely one joule as before:

$$W_{ab_{L1}} = \int_1^2 dx = 1 \tag{5.1-14}$$

This is not the case along the path $L_2$ shown in Figure 5.1-1(b). Since $L_2[1 \le x \le 2, y = 1-(3-2x)^2, z=0]$, $L_2$ reaches the apogee of its trajectory at $x = 1\frac{1}{2}$, $y=1$ as $\overline{F}_2$, which varies as $\hat{u}_x(1-y)$, decreases linearly from a value of one at $y=0$ to a value of zero at $y=1$. The differential work is $\overline{F}_2 \cdot \overline{d\ell} = (1-y)dx$. Our line integral of the form of Eq. (5.1-1) then reduces to the scalar integral

$$W_{ab_{L2}} = \int_{L_2} (1-y)dx \tag{5.1-15}$$

over the path $L_2$.

There are two approaches to integrating Eq. (5.1-15), since the integrand must either be of the form $f(x)dx$ or $g(y)dy$. In this example, the first one is the more straightforward for three reasons. First, $f(x)$ is single-valued over the range from $a$ to $b$. Secondly, $y$ is explicitly given in terms of $x$, namely $y = 1-(3-2x)^2$. Finally, the limits are given over the $dx$ integration from $a$ to $b$ as 1 to 2, respectively. Thus, $W_{ab_{L_2}}$ becomes

$$W_{ab_{L2}} = \int_1^2 (3-2x)^2 dx = \tfrac{1}{3} \text{ joule} \tag{5.1-16}$$

which is clearly different from Eq. (5.1-13).

The other way to evaluate Eq. (5.1-15) is more involved, but is given here for its instructional value. This approach is to solve for $W_{ab_{L_2}}$ by the use of the second form of the integrand, namely $g(y)dy$. There are three steps to this approach. The first is to realize that $g(y)$ is double-valued. Therefore, the $y$ integration must be broken into two regions—one for the rising part of the path (region I) and the other for the falling part (region II). The second is to determine $g(y)$. The third is to determine the $y$ limits for each of the regional integrations.

Here,

$$g(y) = \frac{3 \mp \sqrt{1-y}}{2}$$

where the upper and lower signs are used for region I and region II, respectively. Taking the differential of our path $L_2$ equation and solving for $dx$ yields $dx = +dy/(4\sqrt{1-y})$ for region I, and $dx = -dy/(4\sqrt{1-y})$ for region II. Thus, the two integrations become

$$W_{ab_{L_2}} = \tfrac{1}{4}\int_{0\,I}^{1} \sqrt{1-y}\; dy \; - \tfrac{1}{4}\int_{1\,II}^{0} \sqrt{1-y}\; dy = \tfrac{1}{6} + \tfrac{1}{6} = \tfrac{1}{3}\,\text{joule} \qquad (5.1\text{-}17)$$

which, of course, yields the same answer as Eq. (5.1-16) as it must, since we solved the same problem by two different methods. In the first integral, the $y$ limits are 0 and 1 corresponding to $x = 1$ and 3/2, respectively. In the second integral, the $y$ limits are 1 and 0 corresponding to $x = 3/2$ and 2, respectively.

### 5.1.2(b) Electrostatics—electric field intensity and electric potential

#### *Electric field intensity*

An electric vector force field $\overline{F}$ exists on an isolated test charge $Q_t$ as a result of the presence of a nearby system of electric charges. The force acting on $Q_t$ is proportional to the magnitude of $Q_t$. Electric field intensity, commonly denoted by the symbol $\overline{E}$ is also a force field except that it is referenced to a test charge upon which the force is acting so that it does not depend on the magnitude of the charge upon which it acts. This vector field is called the *electric field intensity*. In fact, the electric field intensity is present even if the test charge is not. It has units of newtons per coulomb in SI units, which is equivalent to volts per meter. By this definition the electric field intensity is

$$\overline{E} = \frac{\overline{F}}{Q_t} \quad \text{newtons per coulomb.} \qquad (5.1\text{-}18)$$

#### *Electric potential*

If an electric field exists, then the differential work done per unit test charge by the electric field on the test charge when the test charge is displaced by the differential amount and direction of $\overline{d\ell}$ is, by definition, the differential electric potential $dV$ , and is determined by

$$dV = \frac{dW}{Q_t} = \frac{\overline{F} \cdot \overline{d\ell}}{Q_t} = \overline{E} \cdot \overline{d\ell} \qquad (5.1\text{-}19)$$

In allowing the test charge to move from point $a$ to point $b$ on $L$, the work per unit charge is

$$V_{ab} = \frac{W_{ab}}{Q_t} = \int_{a\,L}^{b} \overline{E}(\overline{r}) \cdot \overline{d\ell} \qquad (5.1\text{-}20)$$

N·m/C or V. However, the work done by the forces of the field is equal to the *decrease* in potential energy of field. In the conventional definition of electric potential, the quantity $dV$ represents the differential *increase* in potential energy per unit charge done *by* an external forcing system *on* the test charge against the forces of the field. Therefore, the sign is reversed and thus, $dV$ becomes $-\overline{E} \cdot \overline{d\ell}$ and

$$V = -\int \overline{E} \cdot \overline{d\ell} + C \qquad (5.1\text{-}21)$$

where $C$ represents the constant of integration of the indefinite integral [Eq. (5.1-21)].

### 5.1.2(c) Path dependence of tangential line integrals

One key point of the examples in Part (a) was to point out that for some fields the tangential line integral [Eq. (5.1-11)] is independent of the path taken and for others it is not. The general way to determine what vector-field line integrals are independent of the path is to ascertain whether the field is conservative. This is readily done by noting whether the curl of $F$ is zero. Using $\overline{F}_1$ and $\overline{F}_2$ from Part (a), we have $\nabla \times F_1 = 0$. Therefore, $\overline{F}_1$ is deemed conservative and its tangential line integral is path-independent. Further, since $\nabla \times F_2 \neq 0$ , $\overline{F}_2$ is *rotational* and not conservative, and its tangential line integral depends upon the path taken.

Thus far we have confined our discussion to *open line integrals*. Open line integrals are bounded by two points that designate the terminus at each end of the line over which the integration takes place. These two points $a$ and $b$ must not be the same in order for the line integral to be considered open. Although there can be coincident intermediate points, the line is considered open as long as the end points are different. If the end points are at the same position in space, the line integral is said to be closed.

We will discuss the *closed line integral* of Eq. (2.4-21) (tangential form) in the context of Stokes' theorem in Section 5.4. Whereas Stokes' theorem deals with a closed line integral of the form $\oint \overline{F} \cdot \overline{d\ell}$, a special case of expression (5.1-1), closed line integrals may take on any of the forms (5.1-1) through (5.1-10).

### 5.1.3 Other line integral examples

Two additional examples of line integral forms are given in this section. These deal with line-integral forms (5.1-2) and (5.1-3), which are also common in fields and photonics. Although both of these forms yield vector field resultants, the integrands are quite different. One deals with a scalar field $f(\overline{r})$ along a path *L*. The other deals with a vector field $\overline{F}$ along *L* through a cross-product operation. These examples provide different vector fields arising from a filamentary electric current source.

*Example of form (5.1-2): Magnetic vector potential from a filamentary current source*

In this case we let the scalar field in Eq. (5.1-2) be

$$f(\overline{r}\,') = \frac{\mu I(\overline{r}\,')}{4\pi R} \tag{5.1-22}$$

Then the magnetic vector potential $\overline{A}(\overline{r})$ takes the form of expression (5.1-2) as

$$\overline{A}(\overline{r}) = \int_L \frac{\mu I(\overline{r}\,')}{4\pi R} \overline{d\ell}\,' \tag{5.1-23}$$

where the vector potential is determined at the field point $\overline{r}$ due to an electric current source $I(\overline{r}\,')$ on $\overline{d\ell}\,'$ along the filamentary line path *L* designated by the source point at $\overline{r}\,'$. Here μ is the magnetic permeability of the surrounding space and $R = |\overline{r} - \overline{r}\,'|$ is the distance from the field point at $\overline{r}$ to the source point at $\overline{r}\,'$ in $\overline{r}$-space notation [Section 1.1.4].

*Example of form (5.1-3): Magnetic field intensity from a filamentary current source*

In this case we let the vector field in expression (5.1-3) be

$$\overline{F}(\overline{r}\,') = -\frac{I(\overline{r}\,')}{4\pi R^2}\hat{u}_R \tag{5.1-24}$$

Then the magnetic field intensity $\overline{H}(\overline{r})$ takes the form of expression (5.1-3) as

$$\overline{H}(\overline{r}) = -\int_L \frac{I(\overline{r}\,')}{4\pi R^2}\hat{u}_R \times \overline{d\ell\,'} \tag{5.1-25}$$

This is frequently portrayed as the Biot-Savart law that describes the magnetic field intensity $\overline{H}$ resulting from a filamentary electric current source $I(\overline{r}\,')$ on $\overline{d\ell\,'}$ along the filamentary line path $L$ designated by the source point at $\overline{r}\,'$. Here $\hat{u}_R$ is a unit vector in the $\overline{R}$ direction, where $\overline{R} = \overline{r} - \overline{r}\,'$ in $\overline{r}$-space notation and $R^2 = \left|\overline{R}\right|^2$.

## 5.2 Surface Integrals of Vector (and Other Tensor) Fields

As was the case with line integrals, surface integrals were introduced in Section 2.4 to illustrate further examples of the dot-product operation—in that case, of a vector field with vector surface differentials in the integrand of an integral operation. In this section surface integrals are discussed in greater detail. Whereas $\int_S \overline{F}\cdot\overline{da}$ [Eq. (2.4-23)] is the specific surface integral of the orthogonal component of $\overline{F}$ over the surface $A$, we will discuss other surface integrals as well.

### 5.2.1 Surface integrals of scalar, vector and other tensor fields with dot-, cross-, and direct-product integrands

A general definition of a surface integral:

> *An integral of a field quantity taken over a vector differential area $\overline{da}$ that is everywhere normal to a general surface S in space is a surface integral.*

Again, the word "normal" in this definition refers to the surface segment $\overline{da}$ and not to the field, the latter of which may have general direction and be of general rank. In this way, any of the three product operators—dot, cross, or direct—may be applied in the integrand, and any field for which these operations are defined is applicable regardless of rank.

Surface integrals are similar to line integrals in that they have integrands that consist of any of these three operations between scalar, vector or tensor fields, in general, and a differential vector. The principal difference is in the differential vector. Whereas the line-integral differential vector $\overline{d\ell}$ is one-dimensional, the surface-integral differential vector $\overline{da}$ in the integrand is two-dimensional. Thus, in a coordinate system, the surface integrals are double integrals and the differential $\overline{da} = \hat{u}_k d\ell_i d\ell_j$ is a double differential, where $\hat{u}_k = \left(\overline{d\ell_i} \times \overline{d\ell_j}\right) / \left|\overline{d\ell_i} \times \overline{d\ell_j}\right|$. There are numerous forms of surface integrals as there are for line integrals.

In fact, the same ten forms given in Eqs. (5.1-1) through (5.1-10) can represent the different forms of surface integrals. By replacing the $\overline{d\ell}$ s with $\overline{da}$ s and the $L$s with $S$s in Eqs. (5.1-1) through (5.1-10), we have a similar compendium of surface integral forms as we had with line-integral forms in the previous section, where the $S$ designation in the integral denotes a surface in 3D space.

$$\int_S \overline{F}(\overline{r}) \cdot \overline{da} \qquad \text{scalar, } n_R = 0 \qquad (5.2\text{-}1)$$

$$\int_S f(\overline{r})\overline{da} \equiv \int_S f(\overline{r}) \otimes \overline{da} \qquad \text{vector, } n_R = 1 \qquad (5.2\text{-}2)$$

$$\int_S \overline{F}(\overline{r}) \times \overline{da} \qquad \text{vector, } n_R = 1 \qquad (5.2\text{-}3)$$

$$\int_S \overline{\overline{A}}(\overline{r}) \cdot \overline{da} \qquad \text{vector, } n_R = 1 \qquad (5.2\text{-}4)$$

$$\int_S \overline{F}(\overline{r})\overline{da} \equiv \int_S \overline{F}(\overline{r}) \otimes \overline{da} \qquad \text{dyadic, } n_R = 2 \qquad (5.2\text{-}5)$$

$$\int_S \overline{\overline{A}}(\overline{r}) \times \overline{da} \qquad \text{dyadic, } n_R = 2 \qquad (5.2\text{-}6)$$

$$\int_S \overline{\overline{\overline{T}}}(\overline{r}) \cdot \overline{da} \qquad \text{dyadic, } n_R = 2 \qquad (5.2\text{-}7)$$

This process continues, but in general we can summarize all surface-integral forms by taking the dot- cross- and (tensor-) direct-product operations of general rank tensors as we did with line integrals in the prior section.

$$\int_S [{}_{n_R}T_{\mathbf{O}}(\bar{r})]\cdot\overline{da} \quad \Rightarrow \quad {}_{n_R-1}T_{\mathbf{R}} \tag{5.2-8}$$

$$\int_S [{}_{n_R}T_{\mathbf{O}}(\bar{r})]\times\overline{da} \quad \Rightarrow \quad {}_{n_R}T_{\mathbf{R}} \tag{5.2-9}$$

$$\int_S [{}_{n_R}T_{\mathbf{O}}(\bar{r})]\otimes\overline{da} \quad \Rightarrow \quad {}_{n_R+1}T_{\mathbf{R}} \tag{5.2-10}$$

where the post-subscripts "**O**" and "**R**" denote the *operand* and *resultant* tensors as before.

Again, as with the tangential line integral, the most commonly used surface integral form is Eq. (5.2-1), namely $\int_S \bar{F}(\bar{r})\cdot\overline{da}$, except that the component of $\bar{F}$ that is summed is the component that is normal to the surface. Since the $\overline{da}$ in the integrand of forms (5.2-1) through (5.2-10) and the $S$ as the integral region designators are used only for surfaces, contemporary practice is to use the single integral symbol until coordinates and their respective limits of integration are specified. For example, in Cartesian coordinates $(\bar{r}) = (x, y, z)$ and if the surface is parallel to the $xy$-plane, $\overline{da} = \hat{u}_z dxdy$, and Eq. (5.2-1) would be written

$$\int_{y_1}^{y_2}\int_{x_1}^{x_2} \bar{F}(x,y,z)\cdot\hat{u}_z dxdy \tag{5.2-11}$$

where the outside integral goes with the outside differential and the inside integral with the inside differential.

### 5.2.2 Surface integral applications

In Section 4.4 the concept of vector flux was introduced. The flux $\Psi_F$ of a vector field $\bar{F}$ over a surface in space $S$ was found by taking the dot product of each vector surface element $\overline{da}$ with the vector field evaluated at each element over the surface. This process is given by 4.4-2 (also 2.4-26) as

$$\Psi_F = \int_S \bar{F}\cdot\overline{da} \tag{5.2-12}$$

For example, if the vector field is electric current density $\bar{J}$ in amps per square meter, the current $I$ passing through $S$ is the flux of $\bar{J}$ given by

$$I = \int_S \bar{J}\cdot\overline{da} \tag{5.2-13}$$

in amps. Further, if the vector field is electric flux density $\overline{D}$ in coulombs per square meter, the electric flux $\Psi_e$ passing through $S$ is the flux of $\overline{D}$ given by

$$\Psi_e = \int_S \overline{D} \cdot \overline{da} \tag{5.2-14}$$

in coulombs. Likewise, the magnetic flux $\Psi_m$ through $S$ is given by

$$\Psi_m = \int_S \overline{B} \cdot \overline{da} \tag{5.2-15}$$

where $\overline{B}$ is the magnetic flux density in webers per square meter and $\Psi_m$ is the magnetic flux in webers.

Yet another example is the energy flow (flux) through a surface. The instantaneous power density of an electromagnetic wave is given by the Poynting vector $\overline{P} = \overline{E} \times \overline{H}$, where $\overline{E}$ is the electric field intensity (volts per meter) and $\overline{H}$ is the magnetic field intensity (amps per meter), the power density $\overline{P}$ is in watts per $m^2$. The energy flow is then

$$W = \int_S \overline{P} \cdot \overline{da} \tag{5.2-16}$$

watts of power through a surface $S$.

## 5.3 Gauss' (Divergence) Theorem

German mathematician and physicist Johann Karl Friedrich Gauss (1777–1855) first developed and proved a theorem that is a mathematical statement that the density of matter in a volumetric region of space can change only if it flows out of or into the region through its boundary that encloses the volume. This concept was touched upon in Section 4.4.1(c) in the discussion of the significance of zero and nonzero divergence. This theorem has become a useful tool in converting volume integrals of densities of quantities into closed surface integrals of the fluxes of those quantities.

As a precursor to the development of the *divergence theorem*, also known as *Gauss' theorem*, we first discuss another major contribution of Gauss known as *Gauss' law* (Section 5.3.1), which states that the total outward flux of a vector quantity is equal to the total quantity of the enclosed source of that vector field. This leads into the divergence theorem of Gauss (Section 5.3.2).

### 5.3.1 Gauss' law

In the definition of divergence, the total outward flux through a closed surface was needed, as given by Eq. (4.4-1). This expression involved the limit of the ratio of that total outward flux $\Psi_{e_{total}}$ to the enclosed volume as the volume was shrunk to zero about the point in space at which the divergence was to be determined. The numerator of that ratio (namely, the closed-surface integral) has another important interpretation, for example, in Gauss' law:

> *Gauss' law for electrostatics states that the total electric flux emanating outwardly through a closed surface is equal to the total charge enclosed within.*

Here the closed surface and enclosed volume refer to any volume and associated surface, not just the limiting volume in Eq. (4.4-1). Mathematically, *Gauss' law* is expressed as

$$\oint_{S_o} \overline{D} \cdot \overline{da} = Q_{total} \qquad (5.3\text{-}1)$$

where $\overline{D}$ is the electric flux density in coulombs per square meter, and $S_o$ is a generalized closed surface called a *Gaussian surface.*

We know from the above discussion that the left-hand side of Eq. (5.3-1) is also

$$\oint_{S_o} \overline{D} \cdot \overline{da} = \Psi_{e_{total}} \qquad (5.3\text{-}2)$$

where $\Psi_{e_{total}}$ is the total outward electric flux in coulombs. Further, the total charge enclosed within the volume v is

$$Q = \int_{v} \rho dv \qquad (5.3\text{-}3)$$

where the volume integral $\int_{v} \rho dv$ is taken over any volume v enclosed by the Gaussian surface of all charges represented by the electric charge density $\rho$ in $C/m^3$. Eq. (5.3-3) is a generalization of Eq. (4.4-13) for any volume. The result of the volume integral, then, is the total charge $Q$ enclosed within the Gaussian surface, which, of course, has units of coulombs of charge. Thus, the total flux is the total charge

$$\Psi_{e_{total}} = Q_{total} \tag{5.3-4}$$

Therefore, another mathematical statement of *Gauss' law* is

$$\oint_{S_o} \overline{D} \cdot \overline{da} = \int_{\text{v}} \rho d\text{v} \tag{5.3-5}$$

Written in this way, the volume charge density $\rho$ includes point charges $Q_n$ (where $n$ is summed over all point charges), line charges (densities $\rho_L$), surface charges (densities $\rho_s$), as well as any volumes that contain distribution of charges (densities $\rho_{\text{v}}$). Thus,

$$\int_{\text{v}} \rho d\text{v} = \sum_{n=1}^{N_p} Q_n + \sum_{i=1}^{N_\ell} \int_{\ell_i} \rho_{\ell_i} d\ell_i + \sum_{j=1}^{N_s} \int_{s_j} \rho_{s_j} ds_j + \sum_{k=1}^{N_{\text{v}}} \int_{s_k} \rho_{\text{v}_\text{k}} d\text{v}_\text{k} \tag{5.3-6}$$

where all point charges $N_p$, line charges $N_\ell$, surface charges $N_s$, and volume charges $N_{\text{v}}$ within v are summed and included in the generalized volume charge distribution $\rho$ and integrated over v, which includes all regions in space containing the above charge distributions.

### 5.3.2 Derivation of Gauss' divergence theorem

Equation (5.3-5) is valid for all closed surfaces and corresponding enclosed volumes, provided $\overline{D}$ is continuous in the region. In particular it can be applied to a vanishing small volume $\Delta\text{v}$. Dividing both sides of Eq. (5.3-5) by $\Delta\text{v}$, we have

$$\frac{\oint_{S_o} \overline{D} \cdot \overline{da}}{\Delta\text{v}} = \frac{\int_{\Delta\text{v}} \rho d\text{v}}{\Delta\text{v}} \tag{5.3-7}$$

Taking the limit as $\Delta\text{v} \to 0$, the left side is the divergence of $\overline{D}$ by definition as given in Eq. 4.4-1. The right side is $\rho$. Thus,

$$\nabla \cdot \overline{D} = \rho \tag{5.3-8}$$

Equations (5.3-5) and (5.3-8) represent *Maxwell's equations from Gauss' law in integral and differential forms*, respectively. Equation (5.3-8) is also called *Maxwell's Divergence equation for the electric flux density*. This result was alluded to in Section 4.4.1(c). Substituting Eq. (5.3-8) into Eq. (5.3-5), we have the *divergence theorem*

$$\oint_{S_o} \overline{D} \cdot \overline{da} = \int_{v} \nabla \cdot \overline{D}\, dv \tag{5.3-9}$$

which is valid for all vector fields under the condition that $\overline{D}$ and $\nabla \cdot \overline{D}$ are continuous in the region. This theorem is also known as *Gauss' theorem*, not to be confused with *Gauss' law,* which was given in Eq. (5.3-1).

Proof of the divergence theorem can be argued by subdividing the volume v into small volumes $\Delta v_i$ each bounded by a closed surface $S_i$, where all of the volume v is taken up by a finite but large number of differential volumes. The outward flux of one differential volume is the inward flux to neighboring differential volumes except where there is no neighboring volume. Thus, the flux cancels at all common surfaces (in the interior) and the only remaining outward flux is through the original surface $S_o$. If we multiply and divide the left-hand side of Eq. (5.3-9) by $\Delta v_i$ and sum over *i,* we have

$$\oint_{S_o} \overline{D} \cdot \overline{da} = \lim_{\Delta v_i \to 0} \sum_i \underbrace{\left( \lim_{\Delta v_i \to 0} \frac{\oint_{S_i} \overline{D} \cdot \overline{da}}{\Delta v_i} \right)}_{\nabla \cdot \overline{D}} \Delta v_i \tag{5.3-10}$$

Taking the limit as $\Delta v_i \to 0$, the ratio is the divergence by definition [Eq. (4.4-1)] and the limit of the summation over $\Delta v_i$ is the volume integral. This yields the divergence equation (5.3-9), *quod erat demonstradum (QED).*

### 5.3.3 Implications of the divergence theorem on the source distribution

Section 4.4-1(c) provided a quantitative description of the relationship between source distributions and the divergence. Now that we have added the divergence theorem to our medley of mathematical tools we may provide a quantitative description of this relationship. Although we have approached the divergence theorem from the viewpoint of electrostatics, we next generalize this discussion to examine the implications of the divergence theorem on source distributions.

Let us re-examine Eq. (5.3-1). The total amount of outward flux $\Psi_A$ from the vector field $\overline{A}$ emanating from the closed surface $S_o$ that encloses the volume v is due to a distribution of sources $\rho_A$ in the interior of that volume and is described by

$$\Psi_A = \oint_{\Delta S} \overline{A} \cdot \overline{da} = \int_{\Delta v} \rho_A dv \tag{5.3-11}$$

Applying Eq. (5.3-9) to the surface integral of Eq. (5.3-11), we have

$$\Psi_A = \int_v \nabla \cdot \overline{A}\, dv \tag{5.3-12}$$

The surface integral in Eq. (5.3-11) represents the total outward flux of the vector field $\overline{A}$, whereas the volume integral of Eq. (5.3-12) in conjunction with Eq. (5.3-8) represents the decrease of the source density $\rho_A$. Equation (5.3-9) says that these two quantities are the same. Thus, the divergence is given the interpretation of a source density distribution, such as mass per unit volume or charge per unit volume.

### 5.3.4 Application: The energy in electromagnetic fields—Poynting's theorem

An elegant application of the divergence theorem is in the classical determination of the energy in an electromagnetic field.[2] We begin with Maxwell's curl equations for the electric and magnetic field intensities $\overline{E}$ and $\overline{H}$:

$$\nabla \times \overline{E} = -\frac{\partial \overline{B}}{\partial t} \tag{5.3-13}$$

and

$$\nabla \times \overline{H} = \overline{J}_{cd} + \overline{J}_{cv} + \frac{\partial \overline{D}}{\partial t} \tag{5.3-14}$$

where $\overline{D}$ and $\overline{B}$ are the electric and magnetic flux densities, respectively. Here the usual current density $\overline{J}$ is broken out into conduction and convection current densities, $\overline{J}_{cd}$ and $\overline{J}_{cv}$, respectively, because each has its unique energy. Thus, $\overline{J} = \overline{J}_{cd} + \overline{J}_{cv}$. Taking the dot products of $\overline{H}$ with Eq. (5.3-13) and $\overline{E}$ with Eq. (5.3-14) we have

$$\overline{H} \cdot \nabla \times \overline{E} = -\overline{H} \cdot \frac{\partial \overline{B}}{\partial t} \tag{5.3-15}$$

and

$$\bar{E}\cdot\nabla\times\bar{H}=\bar{E}\cdot\left(\bar{J}_{cd}+\bar{J}_{cv}\right)+\bar{E}\cdot\frac{\partial\bar{D}}{\partial t} \qquad (5.3\text{-}16)$$

Subtracting Eq. (5.3-16) from Eq. (5.3-15) we have

$$\bar{H}\cdot\nabla\times\bar{E}-\bar{E}\cdot\nabla\times\bar{H}=-\bar{H}\cdot\frac{\partial\bar{B}}{\partial t}-\bar{E}\cdot\left(\bar{J}_{cd}+\bar{J}_{cv}\right)-\bar{E}\cdot\frac{\partial\bar{D}}{\partial t} \qquad (5.3\text{-}17)$$

Applying the vector identity

$$\bar{H}\cdot\nabla\times\bar{E}-\bar{E}\cdot\nabla\times\bar{H}=\nabla\cdot\left(\bar{E}\times\bar{H}\right) \qquad (5.3\text{-}18)$$

to the left-hand side of Eq. (5.3-17) and integrating over the volume in which the energy is to be determined yields

$$\int_{\mathrm{v}}\left(\bar{E}\cdot\frac{\partial\bar{D}}{\partial t}+\bar{H}\cdot\frac{\partial\bar{B}}{\partial t}+\bar{E}\cdot\bar{J}\right)d\mathrm{v}=-\int_{\mathrm{v}}\nabla\cdot\left(\bar{E}\times\bar{H}\right)d\mathrm{v}\,. \qquad (5.3\text{-}19)$$

Here we apply the divergence theorem Eq. (5.3-9) to the right-hand side of Eq. (5.3-19) to obtain

$$\int_{\mathrm{v}}\left[\bar{E}\cdot\frac{\partial\bar{D}}{\partial t}+\bar{H}\cdot\frac{\partial\bar{B}}{\partial t}+\bar{E}\cdot\left(\bar{J}_{cd}+\bar{J}_{cv}\right)\right]d\mathrm{v}=-\oint_{S}\left(\bar{E}\times\bar{H}\right)\cdot\overline{da}\,. \qquad (5.3\text{-}20)$$

This is the classical *Poynting's theorem*,[3] and is valid for general media, which may be anisotropic, bianisotropic,[4,5] nonlinear, inhomogeneous and/or time variant. The first two terms on the left-hand side of Eq. (5.3-20) represent the time rate of increase of the stored energy of the electromagnetic field in the volume v. The third term represents energy per unit time lost to heat *(the ohmic power)* in the case of the conduction current, and the energy per unit time required to accelerate charges in the case of convection current.[2]

In the event that the volume has energy sources, either electric charge or current sources, the terms on the left would be negative representing power flow out of the region. *All of the net energy must be supplied externally. Thus the term on the right represents the energy flow into the volume per unit time.* Reversing the sign of the right side signifies outward power flow through the surface $S$ that bounds the volume v. The pointing vector $\bar{P}=\bar{E}\times\bar{H}$ may replace the cross product term on the right-hand integrand yielding the total outward power flow $\oint_S \bar{P}\cdot\overline{da}$ (watts).[2]

## 5.4 Stokes' (Curl) Theorem

Sir George Stokes (1819 – 1903) occupied the University of Cambridge Lucasian Chair of Mathematics for the last 54 years of his life. (Once occupied by Isaac Newton before Stokes and now occupied by Stephen Hawking, this is one of the most prestigious chairs in academe). At his death in 1903, the *London Times* gave this accolade: "Sir G. Stokes was remarkable... for his freedom from all personal ambitions and petty jealousies."[6] The theorem that bears his name is the subject of this section.

In the prior section on the development of Gauss' divergence theorem, we found that Gauss' law was fundamental to the definition of divergence. There we were dealing *prima facie* with conservative fields—where the curl is zero and where nonzero divergence had the interpretation of sources within the region. Here we are dealing with the opposite situation, namely where the curl is nonzero and the divergence is zero. This is the case for example with magnetic fields. Before, Gauss' law lead us to the Divergence theorem. Similarly, we find that *Ampere's circuital law* leads us to *Stokes' theorem.*

### 5.4.1 Ampere's circuital law

In the definition of the curl of a vector field, the circulation of the field was needed, as given by Eqs. (4.5-1), (4.5-4), and (4.5-5). This expression involved the limit of the ratio of that circulation to the area of a surface bounded by the closed path as the enclosed surface was shrunk to zero about the point in space at which the curl was to be determined. The numerator of that ratio, namely the closed-line integral, has another important interpretation in *Ampere's circuital law*:

> *Ampere's circuital law states that the circulation of the magnetic field intensity* $\overline{H}$ *about any closed path is equal to the total electric current passing through any surface bounded by that closed path.*

Here the closed line and the enclosed surface refer to any surface and associated closed line, not just the limiting surface in Eq. (4.5-1). Mathematically, *Ampere's circuital law* is expressed as

$$\oint_{L_o} \overline{H} \cdot \overline{d\ell} = I \qquad (5.4\text{-}1)$$

where $\overline{H}$ is the magnetic field intensity in amps per meter (SI units), and $L_o$ is a generalized closed line over which the circulation is determined.

From the discussion surrounding Eq. (5.2-13), the current is the flux of the current density $\overline{J}$ given by

$$I = \int_{S_o} \overline{J} \cdot \overline{da} \tag{5.4-2}$$

where $\overline{J}$ represents all free current passing through $S_o$ bounded by the closed line $L_o$ of Eq. (5.4-1). Therefore, another mathematical statement of *Ampere's circuital law* is

$$\oint_{L_o} \overline{H} \cdot \overline{d\ell} = \int_{S_o} \overline{J} \cdot \overline{da} \tag{5.4-3}$$

### 5.4.2 Derivation of Stokes' theorem

Equation (5.4-3) is valid for all surfaces $S_o$ bounded by $L_o$. In particular, it can be applied to a vanishing small surface $\Delta s$. Dividing both sides of Eq. (5.4-3) by $\Delta s$, we have

$$\frac{\oint_{\Delta L} \overline{H} \cdot \overline{d\ell}}{\Delta s} = \frac{\int_{\Delta s} \overline{J} \cdot \overline{da}}{\Delta s} \tag{5.4-4}$$

where $\Delta L$ is the closed line that bounds $\Delta s$. Taking the limit as $\Delta s \rightarrow 0$, the left side is the definition of the curl as given in Eq. (4.5-1). The right side is $\overline{J}$. Thus,

$$\nabla \times \overline{H} = \overline{J} \tag{5.4-5}$$

Equations (5.4-3) and (5.4-5) represent Maxwell's magnetostatic equations from Ampere's circuital law in integral and differential forms, respectively. Equation (5.4-5) is also called Maxwell's curl equations for the static magnetic field. The concept of a vector field that has a nonzero curl as having flow lines that tend to curl up was discussed in Section 4.5.2. Such fields are then referred to as rotational, solenoidal, or nonconservative.

Substituting Eq. (5.4-5) into Eq. (5.4-3) we have *Stokes' theorem*

$$\oint_{L_o} \overline{H} \cdot \overline{d\ell} = \int_{S_o} \nabla \times \overline{H} \cdot \overline{da} \tag{5.4-6}$$

which is valid for all vector fields under the condition that $\overline{H}$ and $\nabla\times\overline{H}$ are continuous on the surface $S_o$. This theorem is also known as *the curl theorem* because of its obvious use of the curl vector differential operator.

Proof of Stokes' theorem can be argued by subdividing the surface $S_o$ into small surfaces $\Delta s_i$ each bounded by a closed line $\Delta L_i$, where all of the surface $S_o$ is taken up by a finite but large number of differential surfaces. The line integral of one differential surface is the negative of the line integral to neighboring differential surfaces except where there is no neighboring surface, namely on $L_o$. Thus, the line integral cancels at all common surface borders (in the interior) and the only remaining outward contribution is at the original closed line $L_o$. If we multiply and divide the left-hand side of Eq. (5.4-6) by $\Delta s_i$ and sum over *i,* we have

$$\oint_{L_o}\overline{H}\cdot\overline{d\ell}=\lim_{\Delta s_i\to 0}\sum_i\underbrace{\left(\lim_{\Delta s_i\to 0}\frac{\oint_{\Delta L_i}\overline{H}\cdot\overline{d\ell}}{\Delta s_i}\right)}_{(\nabla\times\overline{H})_i}\Delta s_i\,. \tag{5.4-7}$$

Taking the limit as $\Delta \mathrm{v}_i\to 0$, the ratio in the parentheses is the component of the curl normal to each differential surface $\Delta s_i$ by definition [Eq. (4.5-3)]. The limit of the summation over $\Delta s_i$ is the surface integral of that component of the vector field given by $(\nabla\times\overline{H})_i$. This yields Stokes' theorem [Eq. (5.4-6)], and again, *QED.*

### 5.4.3 Implications of Stokes' theorem

If a vector field is solenoidal, i.e., if it has a nonzero curl over a region in space, then from Stokes' theorem, Eq. (5.4-6), the flux of the curl through any bounded surface is equal to the circulation of the tangential component of the original field about the closed path that bounds the surface. This assumes that the surface and its boundary are piecewise continuous. From Ampere's circuital law [Eq. (5.4-1)] this flux is the total current flowing through the closed path. From Eq. (5.4-2) this is the same current that can be determined by integrating the normal component of current density [Eq. (5.4-2)] over any surface bounded by the closed path.

## 5.5 Green's Mathematics

George Green (1793-1841) was a self-taught English mathematician who became interested in electricity and magnetism. In the course of applying potential theory to this area of interest, he developed several integral identities. In March 1828 he privately published (in the Nottingham Subscription Library with only 51 subscribers) his first and perhaps most important paper, "An Essay on the Applications of Mathematical Analysis to the Theories of Electricity and Magnetism." This paper included his later-to-be-appreciated lemma shown in Fig. 5.5-1.[7,8] In this paper, Green presented not only his lemma, which we now refer to as *Green's theorem*, but also other variations of these identities, which are discussed in this section. Various names have been attached to Green's mathematics. Besides *Green's lemma*, these include *Green's formulas*, *Green's identities*, *Green's first and second theorems (of scalar form)* and *Green's first and second theorems of vector form.*

$$\int d\sigma V \frac{dU}{dw} + \int dxdydz V \delta U = \int d\sigma U \frac{dV}{dw} + \int dxdydz U \delta V$$

**Figure 5.5-1** Green's original lemma

Green is also accredited with the concept of the *potential function*, using the term *potential* to describe gravitational potential from the addition of masses in a system weighted by the magnitude of each mass and inversely with the distance to a point where the gravitational potential is being determined. In this book, potential was first mentioned in Section 2.2 and Fig. 2.2-1(a) as an example of a scalar field, then in Section 4.7.2 in the context of scalar and vector potential, and then again in Section 5.1.2 in the context of (a) work or potential energy from a gravitational or mechanical force field and (b) electrostatic potential in conjunction with electric field intensity.

### 5.5.1 Green's identities

Green's first identity for scalar fields $\Psi(\overline{r})$ and $\Phi(\overline{r})$ is

$$\oint_s \Psi\nabla\Phi \cdot \overline{da} = \int_V (\nabla\Psi \cdot \nabla\Phi + \Psi\nabla^2\Phi) dv \qquad (5.5\text{-}1)$$

which is referred to as the asymmetric scalar form. Green's second identity for scalar fields is

$$\oint_s (\Psi\nabla\Phi - \Phi\nabla\Psi)\cdot\overline{da} = \int_V (\Psi\nabla^2\Phi - \Phi\nabla^2\Psi)\,dv \qquad (5.5\text{-}2)$$

which is the symmetric scalar form for obvious reasons. The first vector Green's identity for vector fields $\overline{A}$ and $\overline{B}$ is

$$\oint_s (\overline{A}\times\nabla\times\overline{B})\cdot\overline{da} = \int_V [(\nabla\times\overline{A})\cdot(\nabla\times\overline{B}) - \overline{A}\cdot\nabla\times\nabla\times\overline{B}]dv \qquad (5.5\text{-}3)$$

which is the asymmetric vector form. The second vector Green's identity is

$$\oint_s (\overline{A}\times\nabla\times\overline{B} - \overline{B}\times\nabla\times\overline{A})\cdot\overline{da} = \int_V [\overline{B}\cdot\nabla\times\nabla\times\overline{A} - \overline{A}\cdot\nabla\times\nabla\times\overline{B}]dv \qquad (5.5\text{-}4)$$

which is the symmetric vector form.

### 5.5.2 Green's function

In Section 4.2 we discussed scalar differential operators and presented some simple differential equations stemming from such operators and discussed their solutions. In Section 4.7.3 we developed a second-order scalar differential operator called the scalar Laplacian. Combining these, we have a frequently used partial differential equation where the operator is $\mathcal{L} = \nabla^2 + k^2$. The differential equation for the unknown scalar function $\psi$ is therefore of the form

$$\mathcal{L}\psi = \nabla^2\psi + k^2\psi = 0\,. \qquad (5.5\text{-}5)$$

This form is the *homogeneous Helmholtz scalar wave equation* where $k^2 = \mu\varepsilon\omega^2$ is the temporal eigenvalue—$k$ being the *radian wave number*, i.e., the number of radians of phase per unit length at the time-harmonic ($e^{j\omega t}$) radian frequency $\omega$ for an infinite plane wave in a linear isotropic homogeneous medium having material parameters of permeability $\mu$ and permittivity $\varepsilon$.*

The partial differential equation known as the *inhomogeneous Helmholtz scalar wave equation* can be conveniently solved by means of the *Green's function*. The Green's function is the response to a unit impulse. It is extensively used in the solution to partial differential equations. The idea is this. The inhomogeneous form of $\mathcal{L}\psi(\overline{r}) = \varphi(\overline{r}\,')$ for the case of Eq. (5.5-5) is

---

* This description of $k$ as a *wave number* is more physical than the often used term propagation constant—reserving the term *propagation constant* for $\gamma$ where $\gamma = \alpha + j\beta$, $\alpha$ is the *attenuation constant*, and $\beta$ is the *phase constant*.[3]

$$L\psi(\bar{r}) = \nabla^2\psi(\bar{r}) + k^2\psi(\bar{r}) = \varphi(\bar{r}'), \qquad (5.5\text{-}6)$$

where $\varphi(\bar{r}')$ is a known source function located at source positions $\bar{r}'$. The solution $\psi(\bar{r})$ located at $\bar{r}$ is found by first finding the solution to

$$LG(\bar{r},\bar{r}') = \nabla^2 G(\bar{r},\bar{r}') + k^2 G(\bar{r},\bar{r}') = \delta(\bar{r}-\bar{r}'), \qquad (5.5\text{-}7)$$

where the inhomogeneity term on the right-hand side of Eq. (5.5-7) is the Dirac delta function that has the properties

$$\delta(\bar{r}-\bar{r}') = \begin{cases} 0 & \text{where } \bar{r} \neq \bar{r}' \\ \infty & \text{where } \bar{r} = \bar{r}' \end{cases} \qquad (5.5\text{-}8)$$

and where

$$\int_{v'} \delta(\bar{r}-\bar{r}')dv' = 1 \qquad (5.5\text{-}9)$$

where the integration over v' includes the source point $\bar{r}'$. For this reason, the Dirac delta function is referred to as a *unit impulse.* It is an impulse because of Eq. (5.5-8) and unit because of Eq. (5.5-9). Once $G$ is determined, the solution to Eq. (5.5-6) is determined by

$$\psi(\bar{r}) = \int_{v'} G(\bar{r},\bar{r}')\varphi(\bar{r}')dv' \qquad (5.5\text{-}10)$$

The point here is that because of the nature of the right-hand side of Eq. (5.5-7) given by Eq. (5.5-8) and Eq. (5.5-9), Eq. (5.5-7) is easier to solve than Eq. (5.5-6). Once $G$ is determined, the product of $G$ with the given source function $\varphi(\bar{r}')$ becomes the integrand of Eq. (5.5-10), which can readily be integrated for the solution $\psi(\bar{r})$.

### 5.5.3 Applications of Green's mathematics

In electromagnetics and structural dynamics, we often deal with a time lag between the cause and effect. That is, a stimulus occurs at one point in space and at a given instance in time that results in a field at another point in space and at a later time. In the case of electromagnetic effects, the field that results from a stimulating source must travel at a finite velocity before its effects are observed at a remote location. A stellar example is a supernova—through high-power telescopes, we can currently observe what actually took place perhaps millions or even billions of years ago. Another example occurs every day when we observe and hear a conversation on television with someone on the other side of the earth, especially when a stationary satellite is used to relay the signal. There is a noticeable time delay between the narrator's question and the responder's answer. In fact we viewers have had to become accustomed to the annoyance because of this time lag.

There are many other examples of this retardation in time between stimuli and the resulting field. The thunder heard from a bolt of lightning is the acoustic response that arrives much later than the observed lightning, because the sound travels at approximately one one-millionth that of the light. Acoustic waves are longitudinal waves in the sense that vibrations of the medium are aligned with the direction of travel through the medium. Both longitudinal (acoustic) waves as well as transverse (shear and torsional) waves can exist in solid materials. Each have independent velocities of travel and, thus, each experience a time retardation between their stimulating event and the observed stress and strain fields downstream from the stimulus.

We will see in the following examples that Green's functions and Green's identities are indispensable in describing the resulting retarded fields for electromagnetic and structural dynamics fields eluded to above.

### 5.5.3(a) Retarded electric scalar potential

The inhomogeneous Helmholtz wave equation for the electric scalar potential $V(\overline{r},t)$ at $\overline{r}$ in 3-space (See Section 1.1.4 and Figure 1.1-1) called the field point and at time $t$, due to sources depicted by the volume charge density $\rho(\overline{r}',t')$ at positions $\overline{r}'$ in space called source points and at an earlier time $t'$, is

$$\nabla^2 V(\overline{r},t) - \mu\varepsilon\frac{\partial^2 V(\overline{r},t)}{\partial t^2} = -\frac{\rho(\overline{r}',t')}{\varepsilon} \tag{5.5-11}$$

where $\mu$ and $\varepsilon$ are the permeability and permittivity of the medium assumed to be homogeneous, i.e., uniform in space and not varying in time. Following the method of RWV[9], let us take the Fourier transform of Eq. (5.5-11):

$$\nabla^2 V_{FT}(\overline{r},\omega) + \omega^2\mu\varepsilon V_{FT}(\overline{r},t) = -\frac{\rho_{FT}(\overline{r}',\omega')}{\varepsilon} \tag{5.5-12}$$

where the transforms are defined by the Fourier integrals

$$V_{FT}(\overline{r},\omega)=\int_{-\infty}^{\infty}V(\overline{r},t)e^{-j\omega t}dt \tag{5.5-13}$$

and

$$\rho_{FT}(\overline{r}',\omega)=\int_{-\infty}^{\infty}\rho(\overline{r}',t')e^{-j\omega t'}dt' \tag{5.5-14}$$

with inverse transforms given by

$$V(\overline{r},t)=\frac{1}{2\pi}\int_{-\infty}^{\infty}V_{FT}(\overline{r},\omega)e^{j\omega t}d\omega \tag{5.5-15}$$

and

$$\rho(\overline{r}',t')=\frac{1}{2\pi}\int_{-\infty}^{\infty}\rho_{FT}(\overline{r}',\omega)e^{j\omega t'}d\omega \tag{5.5-16}$$

A solution to Eq. (5.5-12) may be found by Green's function methods, where the Green's function is the solution to

$$\nabla^2G(\overline{r},\overline{r}')+k^2G(\overline{r},\overline{r}')=-\delta(\overline{r}-\overline{r}') \tag{5.5-17}$$

where $k^2=\omega^2\mu\varepsilon$ and $\delta(\overline{r}-\overline{r}')$ is the Dirac delta function defined in Eq. (5.5-8) and Eq. (5.5-9). The minus sign is used to be consistent with Eq. (5.5-12). RWV[5] shows that the solution to the above differential equation is

$$G(\overline{r},\overline{r}')=\frac{e^{\pm jk|\overline{r}-\overline{r}'|}}{4\pi|\overline{r}-\overline{r}'|} \tag{5.5-18}$$

Taking the symmetrical form of Green's scalar identity [Eq. (5.5-2)], letting $\Psi=G$ and $\Phi=V_{FT}(\overline{r},\omega)$ we have

$$\begin{aligned}&\iiint_{v'}\left[G(\overline{r},\overline{r}')\nabla^2V_{FT}(\overline{r},\omega)-V_{FT}(\overline{r},\omega)\nabla^2G(\overline{r},\overline{r}')\right]dv'\\&=\oint_{s}\left[G(\overline{r},\overline{r}')\nabla V_{FT}(\overline{r},\omega)-V_{FT}(\overline{r},\omega)\nabla G(\overline{r},\overline{r}')\right]\cdot\overline{da}\end{aligned} \tag{5.5-19}$$

We let the surface $S$ go to infinity to include all sources. $G$ and $V$ fall off as $1/R$ and the gradients $G\nabla V$ and $V\nabla G$ fall off as $1/R^2$, whereas $\overline{da}$ increases as $R^2$. Therefore, $G\nabla V\cdot\overline{da}$ and $V\nabla G\cdot\overline{da}$ vary as $1/R$ and therefore the right-hand side of Eq. (5.5-19) vanishes. Equation (5.5-19) then becomes

$$\int\limits_{v'}\left[G(\overline{r},\overline{r}')\nabla^2 V_{FT}(\overline{r},\omega)-V_{FT}(\overline{r},\omega)\nabla^2 G(\overline{r},\overline{r}')\right]dv'=0 \qquad (5.5\text{-}20)$$

From Eq. (5.5-12),

$$\nabla^2 V_{FT}(\overline{r},\omega)=-k^2 V_{FT}(\overline{r},t)-\frac{\rho_{FT}(\overline{r}',\omega)}{\varepsilon} \qquad (5.5\text{-}21)$$

and from Eq. (5.5-17),

$$\nabla^2 G(\overline{r},\overline{r}')=-k^2 G(\overline{r},\overline{r}')-\delta(\overline{r}-\overline{r}') \qquad (5.5\text{-}22)$$

Substituting Eq. (5.5-21) and Eq. (5.5-22) into Eq. (5.5-20), we have

$$\begin{aligned}&-\int\limits_{v'}G(\overline{r},\overline{r}')k^2 V_{FT}(\overline{r}',\omega)\,dv'-\int\limits_{v'}G(\overline{r},\overline{r}')\frac{\rho_{FT}(\overline{r}',\omega)}{\varepsilon}dv'\\&+\int\limits_{v'}V_{FT}(\overline{r}',\omega)k^2 G(\overline{r},\overline{r}')\,dv'+\int\limits_{v'}V_{FT}(\overline{r}',\omega)\delta(\overline{r}-\overline{r}')\,dv'=0\end{aligned} \qquad (5.5\text{-}23)$$

The first and third terms cancel and the fourth term is $V_{FT}(\overline{r},\omega)$ through the properties of the Dirac delta function by Eq. (5.5-9). Thus, Eq. (5.5-23) becomes

$$V_{FT}(\overline{r},\omega)=\int\limits_{v'}G(\overline{r},\overline{r}')\frac{\rho_{FT}(\overline{r}',\omega)}{\varepsilon}dv' \qquad (5.5\text{-}24)$$

Substituting Eq. (5.5-18) for $G$ in Eq. (5.5-24), we have

$$V_{FT}(\overline{r},\omega)=\int\limits_{v'}\frac{\rho_{FT}(\overline{r}',\omega)e^{\pm jkR}}{4\pi\varepsilon R}dv' \qquad (5.5\text{-}25)$$

Applying Eq. (5.5-15), we have the inverse Fourier transform

$$V(\overline{r},t)=\int\limits_{v'}\left[\frac{1}{4\pi\varepsilon R}\int_{-\infty}^{\infty}\frac{\rho_{FT}(\overline{r}',\omega)e^{j(\omega t\pm kR)}}{2\pi}d\omega\right]dv' \qquad (5.5\text{-}26)$$

With a time shift $t' = t \pm k\,R/\omega$, the argument of the exponential of Eq. (5.5-26) becomes $e^{j\omega t'}$. Applying the inverse transform defined by Eq. (5.5-16) the integral in Eq. (5.5-26) becomes

$$\int_{-\infty}^{\infty} \frac{\rho_{FT}(\bar{r}',\omega)e^{j\omega t'}}{2\pi} d\omega \equiv \rho(\bar{r}',t') = \rho(\bar{r}',t \pm kv) \qquad (5.5\text{-}27)$$

and the unknown $V(\bar{r},t)$ is

$$V(\bar{r},t) = \int_{v'} \frac{\rho(\bar{r}',t \pm kv)}{4\pi\varepsilon R} dv' \qquad (5.5\text{-}28)$$

where $v = \omega/k = 1/\sqrt{\mu\varepsilon}$, which is the velocity of propagation of an infinite plane wave in the linear isotropic homogeneous medium having material parameters of permeability $\mu$ and permittivity $\varepsilon$. In free space, of course, $v = c$, the free-space velocity of light. Homogeneity refers to material that is not only uniform in space but also is not varying in time.

Whereas the plus sign in Eqs. (5.5-25)–(5.5-28) does indeed lead to a mathematically valid solution to Eq. (5.5-11), namely the plus sign variation of Eq. (5.5-28), such a solution would imply that the response $V(\bar{r},t)$ at $\bar{r}$ happens in advance of the stimulus $\rho(\bar{r}',t+kv)$ at $\bar{r}'$. Feynman refers to these solutions as *advanced potentials.*[10] Since this is akin to causality, practicalities of the real world compel us toward the solution with the minus sign, which implies that the response takes place at a later time or is retarded in time. Thus, we have the *retarded potential* or the *retarded electric scalar potential*

$$V(\bar{r},t) = \int_{v'} \frac{\rho(\bar{r}',t - kv)}{4\pi\varepsilon R} dv' \qquad (5.5\text{-}29)$$

### 5.5.3(b) Retarded magnetic vector potential

A similar development takes place for the magnetic vector potential. The inhomogeneous vector Helmholtz wave equation for the magnetic vector potential $\bar{A}(\bar{r},t)$ at the field point $\bar{r}$ in 3-space and at time $t$, due to sources depicted by the current density vector field $\bar{J}(\bar{r}',t')$ at positions $\bar{r}'$ in space called source points and at an earlier time $t'$, is

$$\nabla^2 \bar{A}(\bar{r},t) - \mu\varepsilon \frac{\partial^2 \bar{A}(\bar{r},t)}{\partial t^2} = -\mu \bar{J}(\bar{r}',t') \qquad (5.5\text{-}30)$$

The usual procedure for finding the resulting vector field $\bar{A}(\bar{r},t)$ in terms of the causal stimulus vector field $\bar{J}(\bar{r}',t')$ is to reduce Eq. (5.5-30) to Cartesian coordinates and to notice that each of the three components take the same form as Eq. (5.5-11) except that the inhomogeneous term is $\mu\bar{J}(\bar{r}',t')$ instead of $\rho(\bar{r}',t')/\varepsilon$.

However, the case must be argued differently for other coordinate systems because components of the vector Laplacian of Eq. (5.5-30) are not simply the scalar Laplacians of the vector field components as they are in Cartesian coordinates. See the vector Laplacian in cylindrical coordinates [Eq. (4.7-14), for example].

Nevertheless, the power of Green's mathematics again is our panacea. A generalized form of Green's identity [Eq. (5.5-2)] is given by Blokh[11]

$$\begin{aligned}&\oint_S\left[G(\bar{r},\bar{r}')\nabla\bar{A}_{FT}(\bar{r},\omega)-\bar{A}_{FT}(\bar{r},\omega)\nabla G(\bar{r},\bar{r}')\right]\cdot\overline{da}\\&=\iint_{v'}\left[G(\bar{r},\bar{r}')\nabla^2\bar{A}_{FT}(\bar{r},\omega)-\bar{A}_{FT}(\bar{r},\omega)\nabla^2G(\bar{r},\bar{r}')\right]dv'\end{aligned}\tag{5.5-31}$$

Again, we let the surface $S$ go to infinity to include all sources. $G$ and $\bar{A}$ fall off as $1/R$ and the gradients $G\nabla\bar{A}$ and $\bar{A}\nabla G$ fall off as $1/R^2$, whereas $\overline{da}$ increases as $R^2$. Therefore, $G\nabla V\cdot\overline{da}$ and $V\nabla G\cdot\overline{da}$ vary as $1/R$ and therefore the left-hand side of Eq. (5.5-31) vanishes as $R\to\infty$, becoming

$$\iint_{v'}\left[G(\bar{r},\bar{r}')\nabla^2\bar{A}_{FT}(\bar{r},\omega)-\bar{A}_{FT}(\bar{r},\omega)\nabla^2G(\bar{r},\bar{r}')\right]dv'=0\tag{5.5-32}$$

which, by a parallel path from which we developed Eq. (5.5-29), leads to

$$\bar{A}(\bar{r},t)=\int_{v'}\mu\frac{\bar{J}(\bar{r}',t-kv)}{4\pi R}dv'\tag{5.5-33}$$

which is *retarded potential* or the *retarded magnetic vector potential*.

## References

1. Richmond B. McQuistan, *Scalar and vector Fields: A Physical Interpretion*, Wiley, New York (1965).

2. Simon Ramo, John R. Whinnery and Theodore Van Duzer, *Fields and Waves in Communication Electronics*, 3rd ed., Wiley, New York (1994).
3. John Henry Poynting, "Transfer of energy in the electromagnetic field," first published in 1884. See *The Columbia Encyclopedia*, 5th ed. Columbia University Press (1994-5).
4. Mustafa Kuzuoglu and Raj Mittra, "A systematic study of perfectly matched absorbers," in *Frontiers in Electromagnetics*, Douglas H. Werner and Raj Mittra, Eds., IEEE Press, Piscataway, NJ (2000).
5. Frederic Mariotte, Bruno Sauviac and Sergei A. Tretyakov, "Artificial bianisotropic composites," in *Frontiers in Electromagnetics*, Douglas H. Werner and Raj Mittra, Eds., IEEE Press, Piscataway, NJ (2000).
6. *London Times*, Feb. 3, 1903.
7. http://www.nottingham.ac.uk/physics/gg/ Copyright 2001, Penny Gowland., University of Nottingham, Nottingham, NG7 2RD, UK.
8. Article by J. J. O'Connor and E. F. Robertson in http://www-groups.dcs.st-andrews.ac.uk/~history/Mathematicians/Green.html
9. Simon Ramo, John R. Whinnery and Theodore Van Duzer, *Fields and Waves in Communication Electronics*, 3rd ed., Wiley, New York (1994).
10. Richard P. Feynman, *Feynman Lectures in Physics*, Addison Wesley Longman, Reading, MA (1970).
11. V. I. Blokh, *Teoriya Uprugosti* (Ukraine: Harkov Unibersitet Izdatelitvo, 1964).

## Appendix A

# Vector Arithmetics and Applications

As a supplement to the vector arithmetics covered in Section 2.4.1, this appendix serves two purposes. First, the commutative and associative laws of vector addition and subtraction are demonstrated. Secondly, these laws may be used to graphically and mathematically bisect vectors. Other vector arithmetic, such as multiplication and division issues, vector-vector dot, cross, and direct products are covered in Sections 2.4.2 and 2.4.3.

Consider two vectors $\overline{A}$ and $\overline{B}$ as shown

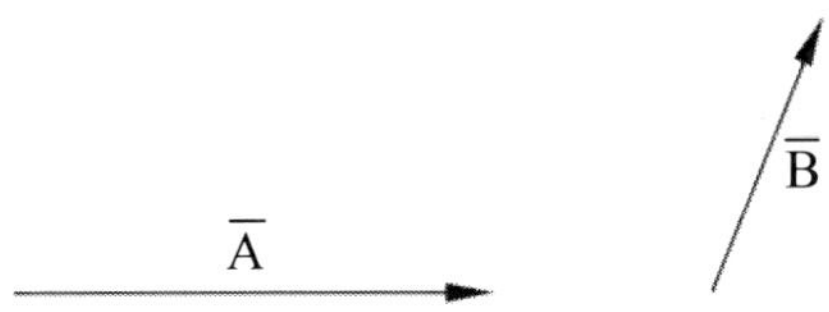


These can be added to yield $\overline{G}$ as shown

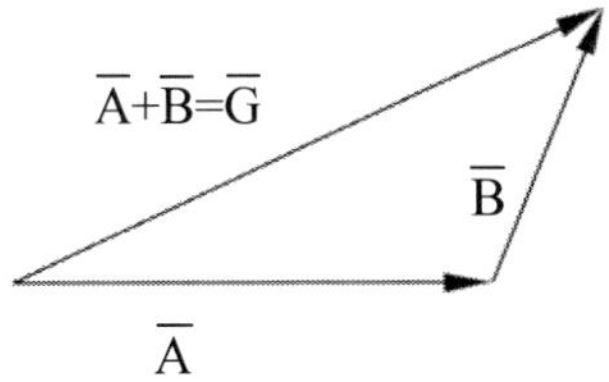


When the order is changed, namely $\overline{A}$ being added to $\overline{B}$, the same vector $\overline{G}$ results

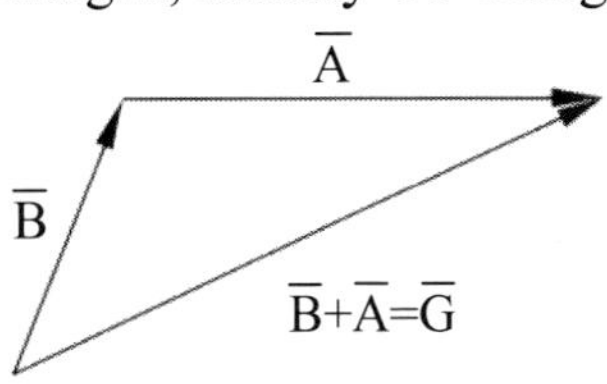

Alternatively $\overline{A}$ and $\overline{B}$ may be added by the following construction

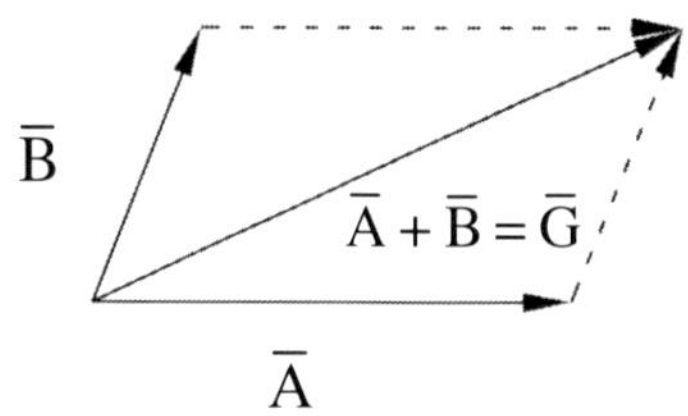


Thus, vector addition of two vectors is *commutative*, namely the sum of vectors is independent of the order in which they are added.

$$\overline{A}+\overline{B}=\overline{B}+\overline{A} \qquad \text{(A-1)}$$

supporting Eq. 2.4-1.

When a third vector $\overline{C}$

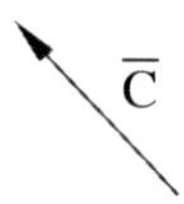


is added to $\overline{A}+\overline{B}$, the resultant vector $\overline{H}=(\overline{A}+\overline{B})+\overline{C}=\overline{G}+\overline{C}$ as shown

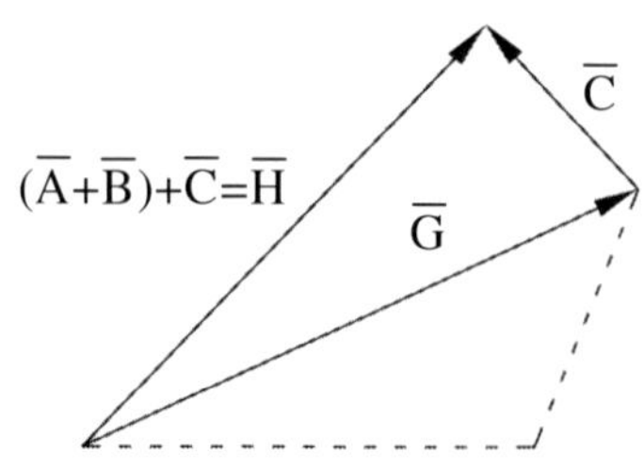


This is the same as if $\overline{C}$ were added to $\overline{B}$ first and then $\overline{A}$ added to the result

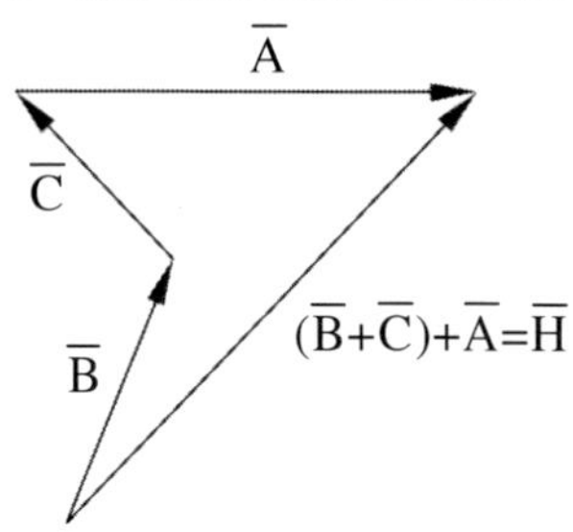

Alternatively, if $\overline{C}$ is added to $\overline{A}$ first and then $\overline{B}$ added to the result, we obtain the same vector $\overline{H}$.

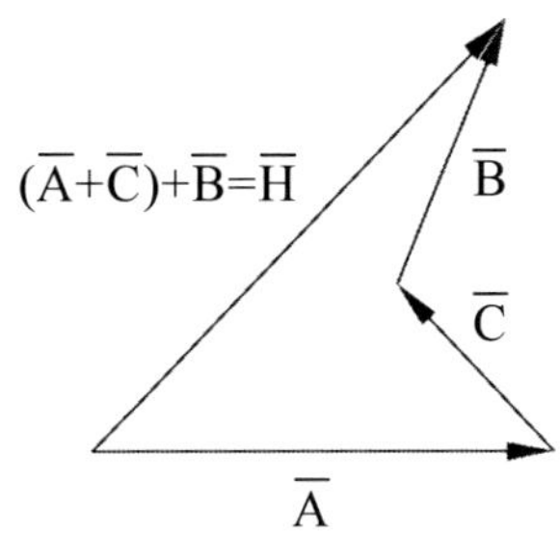


Subtraction works the same. Consider $\overline{A}-\overline{B}=\overline{D}$

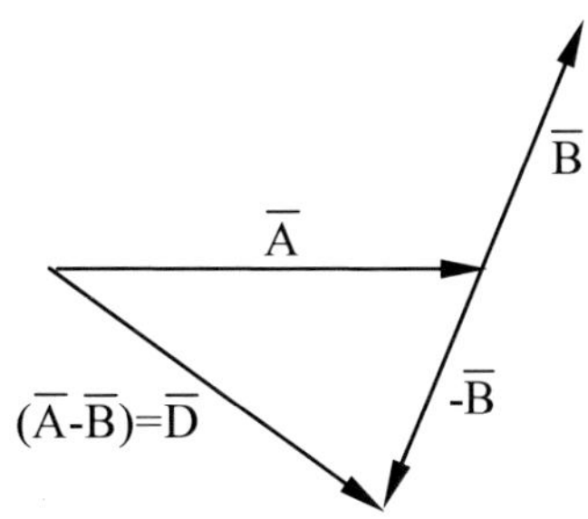


or alternatively, consider $-\overline{B}+\overline{A}$ as

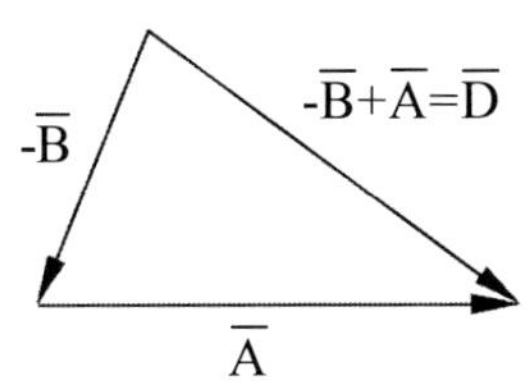


which yields the same result $\overline{D}$

Notice that $\overline{G}+\overline{D}=2\overline{A}$ graphically

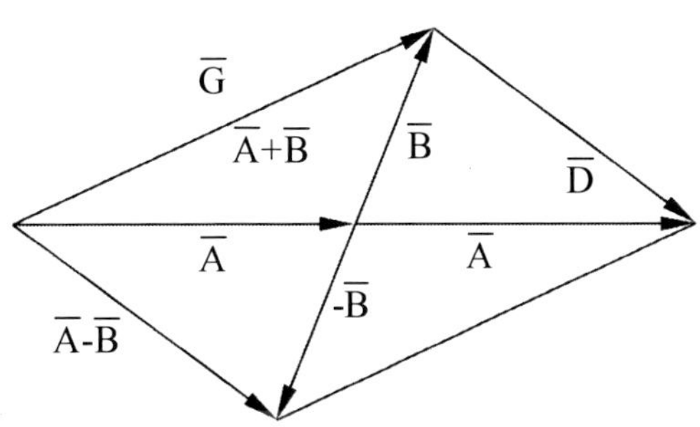

and arithmetically

$$\overline{G} + \overline{D} = \overline{A} + \overline{B} + \overline{A} - \overline{B} = 2\overline{A} \tag{A-2}$$

Also notice that $\overline{G} - \overline{D} = 2\overline{B}$ graphically

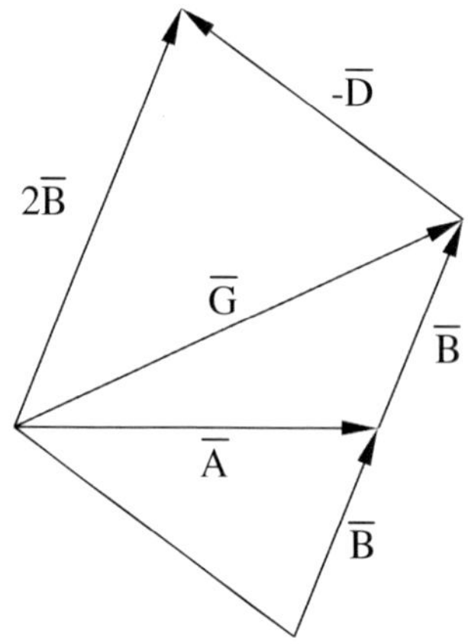


and arithmetically

$$\overline{G} - \overline{D} = \overline{A} + \overline{B} - (\overline{A} - \overline{B}) = 2\overline{B} \tag{A-3}$$

Notice that if we define $\overline{E} = 2\overline{A}$ and $\overline{F} = 2\overline{B}$, the point of the vector $\overline{A}$ *bisects* $\overline{F}$ and $\overline{B}$ *bisects* $\overline{E}$.

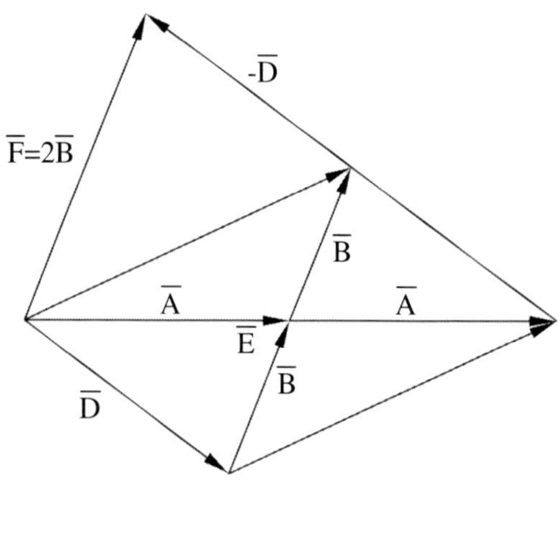

## Appendix B
# Vector Calculus in Orthogonal Coordinate Systems

This appendix first provides a graphical development of first-order vector differential operators in conventional Cartesian and cylindrical coordinates. The focus is in the use made of differential elements of length (DLs) and how DLs become the building blocks for the differential elements of area and volume used in the definitions of gradient, curl, and divergence (and, of course, any combination of these operators). Once this concept is established several tables are presented for detailing the salient parameters associated with other orthogonal coordinate systems. Perspective views of coordinate surfaces are then provided to give graphical views of the coordinate systems.

The *divergence* and *curl* of a vector field are respectively defined in terms of a net outward flux as a differential volume approaches zero and a circulation as three orthogonal differential areas go to zero. These geometries involve the products of orthogonal differential lengths. Therefore, Sections B.1 and B.2 tie differential lengths in Cartesian coordinates to the volumes and areas associated with the divergence and curl, respectively. Section B.3 repeats Section B.1 for cylindrical coordinates as a first step towards curvilinear coordinates from the simplistic Cartesian—cylindrical being the only orthogonal system having only one curvilinear coordinate.

Since the *gradient* is defined in terms of orthogonal differential lengths, it is already cast into an appropriate form for use with the differential-length tables that follow. Section B.4 summarizes the first three sections and provides a description of the geometry for the gradient differential operator.

Section B.5 provides tables of the working parameters associated with these and several other orthogonal coordinate systems.[1-9] Since the prior sections culminate with the expansions for DLs, these tables are therefore focused on the detailed expansion of differential lengths and are presented in each of the orthogonal coordinate systems with their respective transformations to Cartesian coordinates. Such expansions are essential in many practical applications involving vector fields, as well as for other fields regardless of tensor rank.

## B.1 Cartesian Coordinate Geometry for the Divergence

Figure B-1 provides the detailed Cartesian coordinate geometry for the differential volume involved in the mathematical description of the divergence. There are six surfaces making up the differential volume. These are described as Front, Back, Right, Left, Top, and Bottom, as shown in the figure. Each of the differential vector surfaces is defined by its respective corners and normal directions as

$$\overline{da}_i = \hat{u}_i \, d\ell_{i+1} \, d\ell_{i+2} \tag{B.1-1}$$

and the differential volume is

$$d\mathrm{v} = d\ell_i \, d\ell_{i+1} \, d\ell_{i+2} \tag{B.1-2}$$

where $i$=1,2,3, $i$+1=2,3,1 and $i$+2=3,1,2.

In the case of Cartesian coordinates, $i = x,y,z$, $i+1 = y,z,x$ and $i+2 = z,x,y$, respectively. In Fig. B-1 the arbitrary spatial point P is located at $(x,y,z)$, which constitutes the origin "o" of the differential volume involved in the construction of the divergence [Eq. (4.4-1)]. From "o" the three differential lengths $dx$, $dy$ and $dz$ establish the points $a$, $b$ and $c$, respectively. The points diagonal to "o" in the $x$-$y$, $y$-$z$ and $z$-$x$ planes are labeled $d$, $e$, and $f$, respectively, and the point P+dP located at $x+dx$, $y+dy$, $z+dz$ is labeled $g$, to finish out the eight corners of our differential volume $d\mathrm{v}$.

Therefore the Front surface $adgf$ located at $x+dx$ is denoted as $\overline{da}_{adgf} = \hat{u}_x \, dy\, dz\big|_{x+dx}$, where the outward normal is in the positive $x$-direction and $dy\, dz\big|_{x+dx}$ is the magnitude of the area at $x+dx$. (In Cartesian coordinates $dy\, dz$ is invariant, but in all other coordinate systems the differential area may change.) Likewise the Back surface $oceb$ located at $x$, is denoted as $\overline{da}_{oceb} = -\hat{u}_x \, dy\, dz\big|_x$, where the outward normal is in the negative $x$-direction. The remaining four surfaces are similarly constructed yielding $\overline{da}_{begd} = \hat{u}_y \, dx\, dz\big|_{y+dy}$, $\overline{da}_{oafc} = -\hat{u}_y \, dx\, dz\big|_y$, $\overline{da}_{cfge} = \hat{u}_z \, dx\, dy\big|_{z+dz}$, and $\overline{da}_{obda} = -\hat{u}_z \, dx\, dy\big|_z$, for the Right, Left, Top, and Bottom, respectively. These are detailed in Fig. B-1. Finally the differential volume $d\mathrm{v}$ is denoted as $d\mathrm{v} = dx\, dy\, dz$.

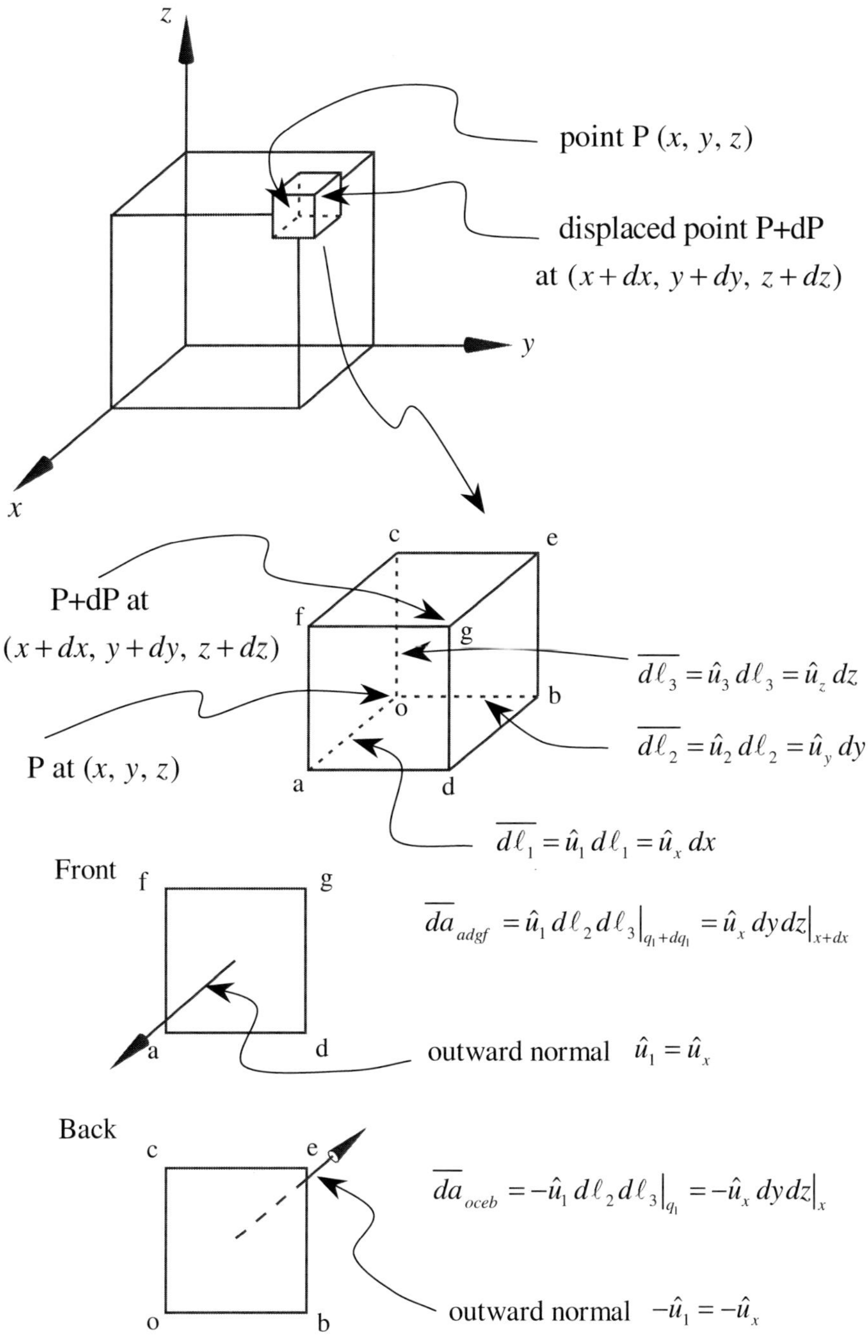


**Figure B-1.** Differential volume in Cartesian coordinates used in the development of the divergence (two pages).

Right

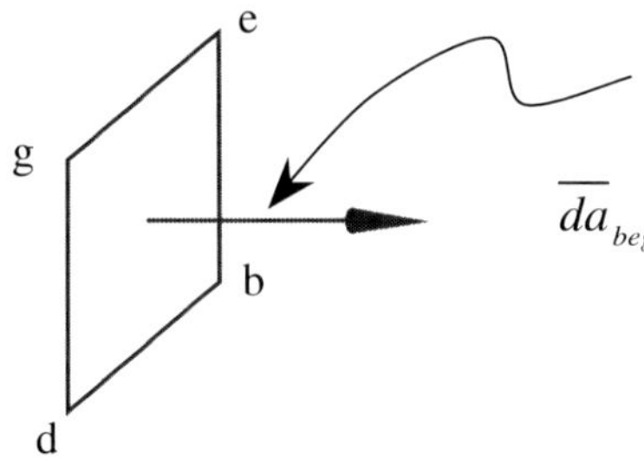


outward normal $\hat{u}_2 = \hat{u}_y$

$$\overline{da}_{begd} = \hat{u}_2\, d\ell_1 d\ell_3\Big|_{q_2+dq_2} = \hat{u}_y\, dxdz\Big|_{y+dy}$$

Left

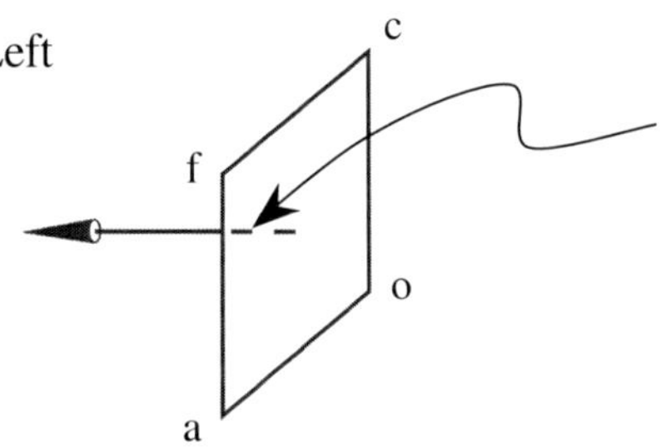


outward normal $-\hat{u}_2 = -\hat{u}_y$

$$\overline{da}_{oafc} = -\hat{u}_2\, d\ell_1 d\ell_3\Big|_{q_2} = -\hat{u}_y\, dxdz\Big|_{y}$$

Top

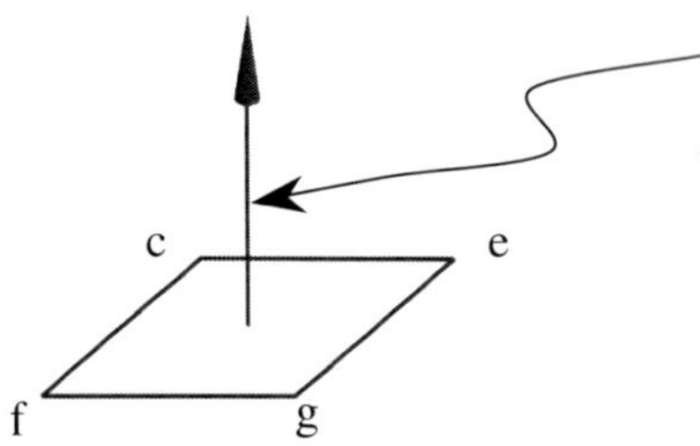


outward normal $\hat{u}_3 = \hat{u}_z$

$$\overline{da}_{cfge} = \hat{u}_3\, d\ell_1 d\ell_2\Big|_{q_3+dq_3} = \hat{u}_z\, dxdy\Big|_{z+dz}$$

Bottom

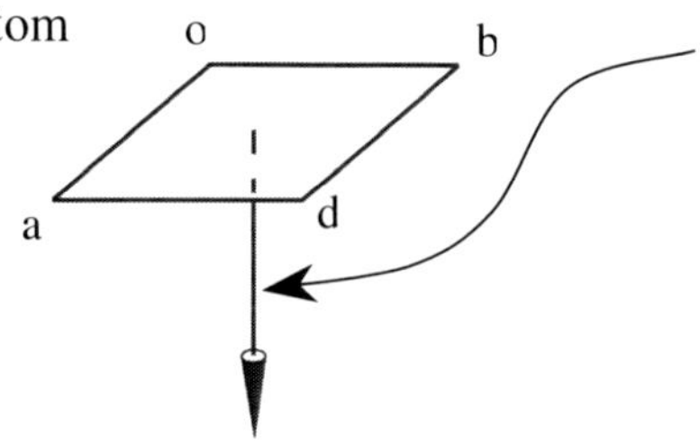


outward normal $-\hat{u}_3 = -\hat{u}_z$

$$\overline{da}_{obda} = -\hat{u}_3\, d\ell_1 d\ell_2\Big|_{q_3} = -\hat{u}_z\, dxdy\Big|_{z}$$

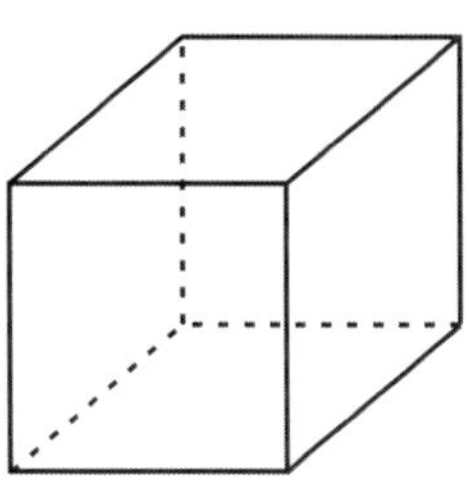

$$dv = d\ell_1 d\ell_2 d\ell_3 = dxdydz$$

Following the procedure of Section 4.4.2 the divergence of a vector field $\overline{A} = \hat{u}_x A_x + \hat{u}_y A_y + \hat{u}_z A_z$ in Cartesian coordinates becomes

$$\nabla \cdot \overline{A}\Big|_{Cartesian} = \frac{\partial A_x}{\partial x} + \frac{\partial A_y}{\partial y} + \frac{\partial A_z}{\partial z} \tag{B.1-3}$$

which is a scalar as expected from Table 4-1, row **D**, column **v**.

Further, following the procedure of Section 4.7.4(a) a dyadic field

$$\begin{aligned} \overline{\overline{G}} = &\ \hat{u}_x\hat{u}_x G_{xx} + \hat{u}_x\hat{u}_y G_{xy} + \hat{u}_x\hat{u}_z G_{xz} \\ &+ \hat{u}_y\hat{u}_x G_{yx} + \hat{u}_y\hat{u}_y G_{yy} + \hat{u}_y\hat{u}_z G_{yz} \\ &+ \hat{u}_z\hat{u}_x G_{zx} + \hat{u}_z\hat{u}_y G_{zy} + \hat{u}_z\hat{u}_z G_{zz} \end{aligned} \tag{B.1-4}$$

has a divergence in Cartesian coordinates given by

$$\begin{aligned} \nabla \cdot \overline{\overline{G}}\Big|_{Cartesian} = &\ \hat{u}_x\left[\frac{\partial G_{xx}}{\partial x} + \frac{\partial G_{yx}}{\partial y} + \frac{\partial G_{zx}}{\partial z}\right] \\ &+ \hat{u}_y\left[\frac{\partial G_{xy}}{\partial x} + \frac{\partial G_{yy}}{\partial y} + \frac{\partial G_{zy}}{\partial z}\right] \\ &+ \hat{u}_z\left[\frac{\partial G_{xz}}{\partial x} + \frac{\partial G_{yz}}{\partial y} + \frac{\partial G_{zz}}{\partial z}\right] \end{aligned} \tag{B.1-5}$$

which is a vector consistent with the rules of Table 4-1, row **D**, column **d**. Equation (B.1-5) follows from Eq. (4.7-9) where $i, j = 1,2,3 = x, y, z$, respectively, all three metric coefficients, $h_1$, $h_2$, and $h_3$, are unity, and all unit vector derivatives in the second term of the square brackets of Eq. (4.7-9) are zero.

## B.2 Cartesian Coordinate Geometry for the Curl

Figure B-2 provides the detailed Cartesian coordinate geometry for the differential surfaces involved in the mathematical description of each of the vector components of the curl [Eq. (4.5-1)]. There are three such orthogonal surfaces used in the definition of the components of the curl. These are the same as the Back, Left, and Bottom surfaces with the same corner labels as in Fig. B-1, except that their vector directions are all positive. As before, each of the differential vector surfaces are defined by their respective corners and normal directions as

$$\overline{da}_{obec} = \overline{da}_x = \hat{u}_x \, dy \, dz\big|_x , \tag{B.2-1}$$

$$\overline{da}_{ocfa} = \overline{da}_y = \hat{u}_y \, dz \, dx\big|_y \tag{B.2-2}$$

and

$$\overline{da}_{oadb} = \overline{da}_z = \hat{u}_z \, dx \, dy\big|_z \tag{B.2-3}$$

From these the curl is determined in Cartesian coordinates following the procedure of Section 4.5.3 and outlined in Fig. B-2.

Following the procedure of Section 4.5.3 the curl of a vector field $\overline{A} = \hat{u}_x A_x + \hat{u}_y A_y + \hat{u}_z A_z$ in Cartesian coordinates becomes

$$\begin{aligned} \nabla \times \overline{A}\Big|_{Cartesian} &= \hat{u}_x \left( \frac{\partial A_z}{\partial y} - \frac{\partial A_y}{\partial z} \right) \\ &+ \hat{u}_y \left( \frac{\partial A_x}{\partial z} - \frac{\partial A_z}{\partial x} \right) \\ &+ \hat{u}_z \left( \frac{\partial A_y}{\partial x} - \frac{\partial A_x}{\partial y} \right) \end{aligned} \tag{B.2-4}$$

which is a vector as expected from Table 4-1, row **C**, column **v**.

Further following the procedure of Section 4.7.4(a) the curl of a dyadic field given by Eq. (B.1-4) in Cartesian coordinates is

$$\begin{aligned} \nabla \times \overline{\overline{G}}\Big|_{Cart} &= \hat{\hat{u}}_{xx} \left[ \frac{\partial G_{zx}}{\partial y} - \frac{\partial G_{yx}}{\partial z} \right] + \hat{\hat{u}}_{xy} \left[ \frac{\partial G_{zy}}{\partial y} - \frac{\partial G_{yy}}{\partial z} \right] + \hat{\hat{u}}_{xz} \left[ \frac{\partial G_{zz}}{\partial y} - \frac{\partial G_{yz}}{\partial z} \right] \\ &+ \hat{\hat{u}}_{yx} \left[ \frac{\partial G_{xx}}{\partial z} - \frac{\partial G_{zx}}{\partial x} \right] + \hat{\hat{u}}_{yy} \left[ \frac{\partial G_{xy}}{\partial z} - \frac{\partial G_{zy}}{\partial x} \right] + \hat{\hat{u}}_{yz} \left[ \frac{\partial G_{xz}}{\partial z} - \frac{\partial G_{zz}}{\partial x} \right] \\ &+ \hat{\hat{u}}_{zx} \left[ \frac{\partial G_{yx}}{\partial x} - \frac{\partial G_{xx}}{\partial y} \right] + \hat{\hat{u}}_{zy} \left[ \frac{\partial G_{yy}}{\partial x} - \frac{\partial G_{xy}}{\partial y} \right] + \hat{\hat{u}}_{zz} \left[ \frac{\partial G_{yz}}{\partial x} - \frac{\partial G_{xz}}{\partial y} \right] \end{aligned} \tag{B.2-5}$$

which is a dyadic consistent with the rules of Table 4-1, row **C**, column **d**. Equation (B.2-5) is a special case of Eq. (C.2-6) in the next appendix for a dyadic (second-rank tensor). An application of Eq. (B.2-5) can be found in Appendix D on page D-2.

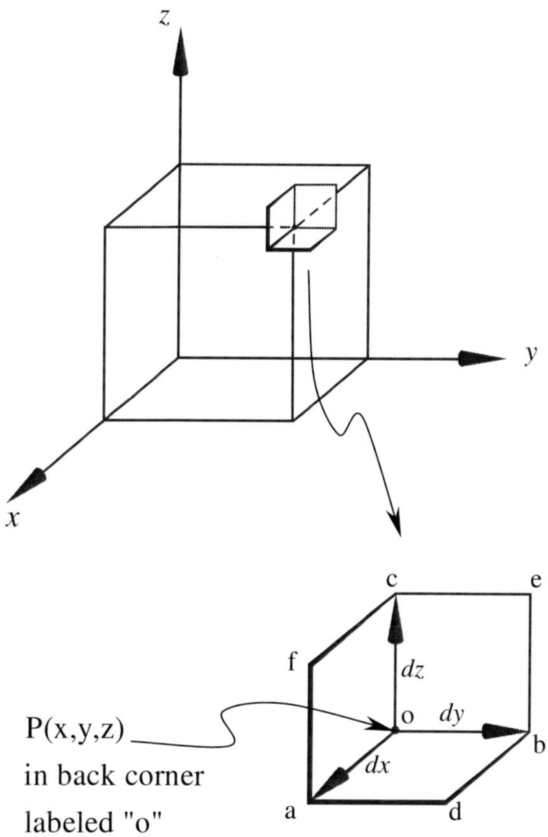


$$\overline{da}_1 = \hat{u}_1 d\ell_2 d\ell_3 = \overline{da}_{obec} = \hat{u}_x\, dy\, dz\Big|_x$$

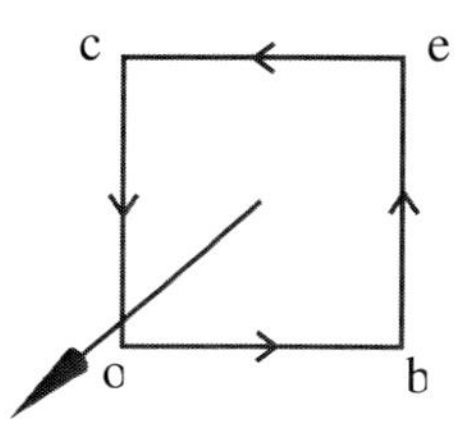


$$\lim_{\Delta a_1 \to 0} \left[ \frac{\int_o^b + \int_b^e + \int_e^c + \int_c^o}{\Delta a_1} \right]$$

$$(\nabla \,\mathrm{x}\, \overline{A})_x = \frac{\partial A_z}{\partial y} - \frac{\partial A_y}{\partial z}$$

$$(\nabla \,\mathrm{x}\, \overline{A})_1 = (\nabla \,\mathrm{x}\, \overline{A})_x$$

**Figure B-2** Differential surfaces in Cartesian coordinates used in the development of the curl (two pages).

$$\overline{da}_2 = \hat{u}_2 d\ell_1 d\ell_3 = \overline{da}_{ocfa} = \hat{u}_y dxdz\Big|_y$$

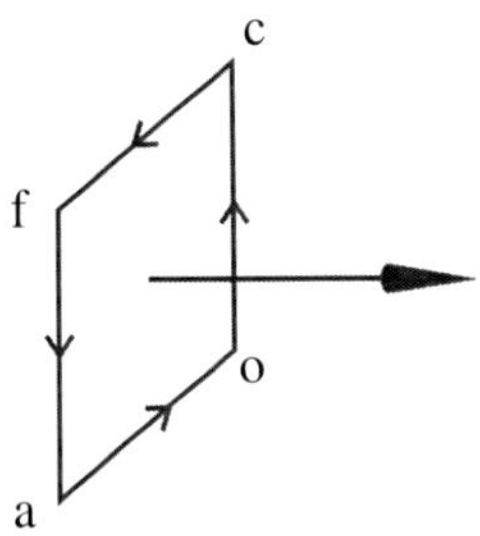


$$\lim_{\Delta a_2 \to 0} \left[ \frac{\int_o^c + \int_c^f + \int_f^a + \int_a^o}{\Delta a_2} \right]$$

$$(\nabla \text{ x } \overline{A})_y = \frac{\partial A_x}{\partial z} - \frac{\partial A_z}{\partial x}$$

$$(\nabla \text{ x } \overline{A})_2 = (\nabla \text{ x } \overline{A})_y$$

$$\overline{da}_3 = \hat{u}_3 d\ell_1 d\ell_2 = \overline{da}_{oadb} = \hat{u}_z dxdy\Big|_z$$

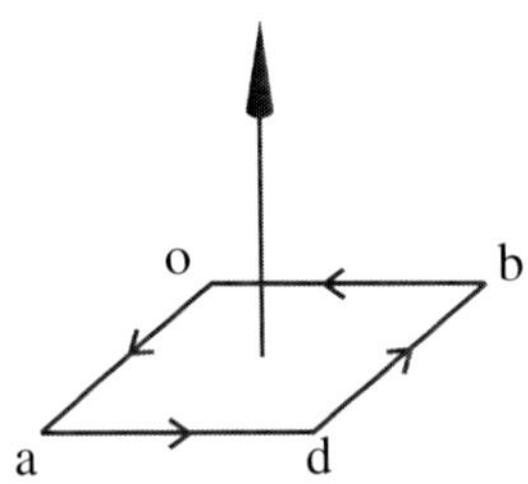


$$\lim_{\Delta a_3 \to 0} \left[ \frac{\int_o^a + \int_a^d + \int_d^b + \int_b^o}{\Delta a_3} \right]$$

$$(\nabla \text{ x } \overline{A})_z = \frac{\partial A_y}{\partial x} - \frac{\partial A_x}{\partial y}$$

$$(\nabla \text{ x } \overline{A})_3 = (\nabla \text{ x } \overline{A})_z$$

## B.3 Cylindrical Coordinate Geometry for the Divergence

In the previous two sections the differential lengths were invariant with changes in coordinate position. This is because the metric coefficients $h_1$, $h_2$, and $h_3$ are unity in the Cartesian system, and, thus, are not functions of the coordinate positions $x,y,z$. However, this simplicity is not the case in any other orthogonal coordinate system. The simplest generalization of this effect is seen in cylindrical coordinates, since the second metric coefficient $h_2$ is not unity, but is given by

$$h_2 = h_\phi = r \qquad \text{(B.3-1)}$$

whereas, $h_1$ and $h_3$ remain unity. In fact it can be stated that

*The circular cylindrical coordinate system is the only non-Cartesian orthogonal system having only one curvilinear coordinate, namely the azimuthal coordinate $\phi$.*

Figure B-3 repeats the development of Fig. B-1 with the effects of Eq. (B.3-1) carefully taken into account in its development for the cylindrical system. In Fig. B-3 the arbitrary spatial point P is located at $(r,\phi,z)$, which constitutes the origin "o" of the differential volume involved in the construction of the divergence from Eq. (4.4-1). From "o" the three differential lengths $d\ell_1, d\ell_2$, and $d\ell_3$ are $dr$, $r\,d\phi$, and $dz$, respectively. These establish the points $a$, $b$, and $c$, respectively, where

$$d\ell_2 = h_2\,dq_2 = r\,d\phi \qquad \text{(B.3-2)}$$

specifies the second differential length. The points diagonal to "o" in the $r$-$\phi$, $\phi$-$z$, and $z$-$r$ planes are labeled $d$, $e$, and $f$, respectively, and the point P+dP located at $r+dr$, $\phi+d\phi$, $z+dz$ is labeled $g$, to finish out the eight corners of $dv$.

Notice that because of Eq. (B.3-2) all six of surfaces are affected by Eq. (B.3-1); however, two pairs of surfaces have the same area and one pair does not. Notice also that two pairs are rectangles and one pair is not. The pair that does not have the same area is the $r$ coordinate pair as seen by

$$\overline{da}_{adgf} = \overline{da}_{r^+} = \hat{u}_r\, r\,d\phi\,dz\Big|_{r+dr} \qquad \text{(B.3-3)}$$

which is labeled as the Front in Fig. B-3, and

$$\overline{da}_{oceb} = \overline{da}_{r^-} = -\hat{u}_r\, r\, d\phi\, dz\Big|_r \qquad \text{(B.3-4)}$$

which is labeled Back, and where $\overline{da}_{r^+}$ and $\overline{da}_{r^-}$ are the positively and negatively directed vector areas at $r$ and $r+dr$, respectively. Although the Front and Back surfaces are rectangles, $\overline{da}_{r^+} > \overline{da}_{r^-}$ because $(r+dr) > dr$.

The other rectangular pair of surfaces is made up of the Left and Right sides. Whereas the scalar areas are equal, their directions are not, as seen by

$$\overline{da}_{oafc} = \overline{da}_{\phi^-} = -\hat{u}_\phi\, dr\, dz\Big|_\phi \qquad \text{(B.3-5)}$$

and

$$\overline{da}_{begd} = \overline{da}_{\phi^+} = \hat{u}_\phi\, dr\, dz\Big|_{\phi+d\phi} \qquad \text{(B.3-6)}$$

because $dr$ and $dz$ are the same at $\phi$ and $\phi+d\phi$, that is, $\left|\overline{da}_{\phi^-}\right| = \left|\overline{da}_{\phi^+}\right|$; but $\overline{da}_{\phi^-}$ at $\phi$ is not collinear with $\overline{da}_{\phi^+}$ at $\phi+d\phi$ as evident from the Fig. B-3.

The remaining pair is the Top and Bottom. These are

$$\overline{da}_{cfge} = \overline{da}_{z^+} = \hat{u}_z\, dr\, r\, d\phi\Big|_{z+dz} \qquad \text{(B.3-7)}$$

and

$$\overline{da}_{obda} = \overline{da}_{z^-} = -\hat{u}_z\, dr\, r\, d\phi\Big|_z \qquad \text{(B.3-8)}$$

which are equal in area but are not rectangles.

Following the procedure of Section 4.4.2 the divergence of a vector field $\overline{A} = \hat{u}_r A_r + \hat{u}_\phi A_\phi + \hat{u}_z A_z$ in cylindrical coordinates becomes

$$\nabla \cdot \overline{A}\Big|_{cyl} = \frac{1}{r}\left[\frac{\partial(rA_r)}{\partial r} + \frac{\partial A_\phi}{\partial \phi} + \frac{\partial(rA_z)}{\partial z}\right] \qquad \text{(B.3-9)}$$

which is a scalar as expected from Table 4-1, row **D**, column **v**.

Further following the procedure of Section 4.7.4(a) the divergence of a dyadic field given by

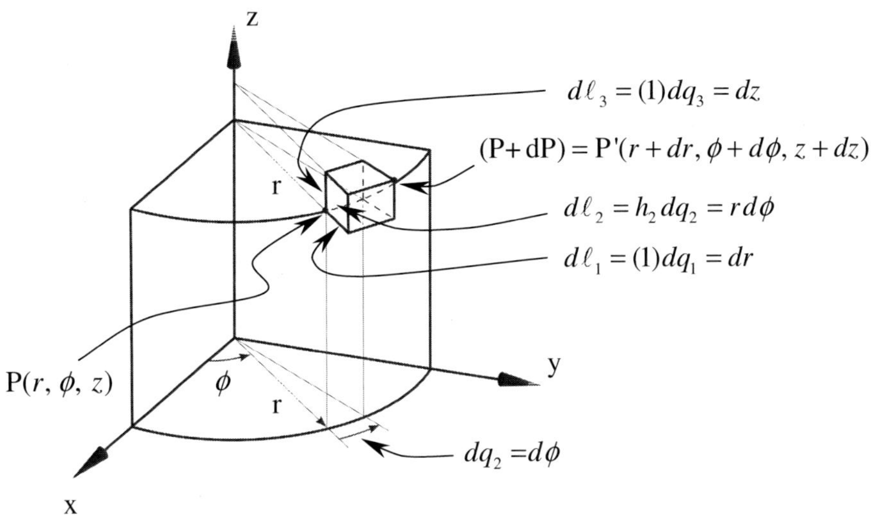


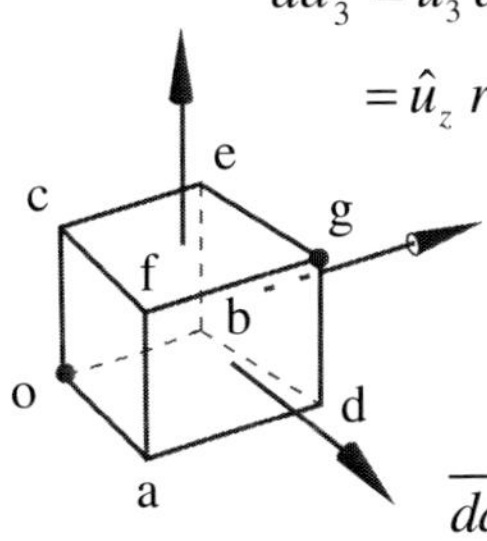


$$\overline{da}_3 = \hat{u}_3\, d\ell_1\, d\ell_2 = \hat{u}_z\, r dr\, d\phi\Big|_{z+dz} \quad \text{Top}$$

$$\overline{da}_2 = \hat{u}_2\, d\ell_1\, d\ell_3 = \hat{u}_\phi\, dr\, dz\Big|_{\phi+d\phi} \quad \text{Right side}$$

$$\overline{da}_1 = \hat{u}_1 d\ell_2\, d\ell_3 = \hat{u}_r\, r d\phi dz\Big|_{r+dr} \quad \text{Front}$$

Back $\quad \overline{da}_1 = -\hat{u}_r\, r d\phi dz\Big|_r$

Left side $\quad \overline{da}_2 = -\hat{u}_\phi\, r dr dz\Big|_\phi$

Bottom $\quad \overline{da}_3 = -\hat{u}_z\, r dr\, d\phi\Big|_z$

$$dv = d\ell_1\, d\ell_2 d\ell_3 = h_1 h_2 h_3 dq_1 dq_2 dq_3 = r dr d\phi dz$$

$$x = r\cos\phi, \quad y = r\sin\phi, \quad z = z$$

**Figure B-3** Geometrical construct for the divergence in circular cylindrical coordinates.

$$\begin{aligned}\bar{\bar{G}} &= \hat{u}_r\hat{u}_r G_{rr} + \hat{u}_r\hat{u}_\phi G_{r\phi} + \hat{u}_r\hat{u}_z G_{rz} \\ &+ \hat{u}_\phi\hat{u}_r G_{\phi r} + \hat{u}_\phi\hat{u}_\phi G_{\phi\phi} + \hat{u}_\phi\hat{u}_z G_{\phi z} \\ &+ \hat{u}_z\hat{u}_r G_{zr} + \hat{u}_z\hat{u}_\phi G_{z\phi} + \hat{u}_z\hat{u}_z G_{zz}\end{aligned} \qquad \text{(B.3-10)}$$

in cylindrical coordinates is

$$\nabla\cdot\bar{\bar{G}}\Big|_{cyl} = \frac{1}{r}\left\{\begin{aligned} &\hat{u}_r\left[\frac{\partial(rG_{rr})}{\partial r} + \frac{\partial G_{\phi r}}{\partial\phi} - G_{\phi\phi} + r\frac{\partial G_{zr}}{\partial z}\right] \\ &+\hat{u}_\phi\left[\frac{\partial(rG_{r\phi})}{\partial r} + \frac{\partial G_{\phi\phi}}{\partial\phi} + G_{\phi r} + r\frac{\partial G_{z\phi}}{\partial z}\right] \\ &+\hat{u}_z\left[\frac{\partial(rG_{rz})}{\partial r} + \frac{\partial G_{\phi z}}{\partial\phi} + r\frac{\partial G_{zz}}{\partial z}\right]\end{aligned}\right\} \qquad \text{(B.3-11)}$$

which is a vector consistent with the rules of Table 4-1, row **D**, column **d**. Equation (B.3-11) follows from Eq. (4.7-9) where $i,j=1,2,3=r,\phi,z$, respectively, with two metric coefficients, $h_1$ and $h_3$ being set to unity and $h_2$ being set to $r$. Further, two of the coordinate unit vectors in the second term of the square brackets of Eq. (4.7-9) are nonzero as given by Eq. (4.3-21) and Eq. (4.3-22) resulting in the two terms of Eq. (B.3-11) that are shown without the partial derivatives, i.e., $G_{\phi r}$ and $G_{\phi\phi}$. Without these terms the divergence would, of course, be incorrect.

*When dealing with any orthogonal coordinate system other than Cartesian, coordinate derivatives of unit vectors must be taken into account in determining vector differential operator expansions.*

## B.4 Summary of the Geometries for Divergence, Curl, and Gradient

In the first three sections of this Appendix we have focused on the geometries of the divergence and curl that require the determination of differential lengths in order to obtain their respective volumes and areas. Since the gradient is already defined in terms of differential lengths, it was not necessary to provide the graphic details of the geometry for this vector differential operator; however, it is described below.

Sections B.1 and B.2 provided Cartesian coordinate geometries for the volume and areas needed for the divergence and curl differential operators. In Section B.1 six surfaces were used for the construction of the *closed surface integral* that made up the differential volume needed for the divergence. In Section B.2 only the back three surfaces were needed. These three surfaces made up the orthogonal bounded surfaces needed for the *closed line integrals* required for the three components of the curl. These geometrical areas and volume were broken down into products of two differential lengths for the surfaces and three differential lengths for the volume.

The geometry for the gradient was considered unnecessary to develop because the three components of gradient would just be derived from three orthogonal differential vectors $\overline{dx}$ from "o" to point $a$, $\overline{dy}$ from "o" to point $b$ and $\overline{dz}$ from "o" to point $c$, referring to Figs. B-1, B-2, or B-3.

Section B.3 provided the cylindrical coordinate geometry for *divergence* as a first step into considerations that come into play when dealing with curvilinear coordinates, since its azimuthal coordinate is curvilinear. In order to obtain the cylindrical coordinate geometry for the *curl* the three open surfaces needed would be *obec* for the $r$-component, *ocfa* for the $\phi$-component and *oadb* for the $z$-component of the curl, referring to Fig. B-3.

In each case the order taken leads to closed line integrals in the direction such that the vector components are positive by the right-hand rule. Finally, we may mentally construct the cylindrical coordinate geometry for the *gradient* by noting that the three components of the gradient in cylindrical coordinates would be derived from three orthogonal differential vectors $\overline{dr}$ from "o" to point $a$, $\overline{rd\phi}$ from "o" to point $b$ and $\overline{dz}$ from "o" to point $c$.

## B.5 Orthogonal Coordinate System Parameters and Surface Graphics

Since the above developments for divergence, curl, and gradient (and all of their combinations) stem from a precise description of differential elements of length, we next provide useful tables of several orthogonal coordinate systems that lead to differential elements of length and other parameters unique to each coordinate system. Finally, we graphically present the orthogonal coordinate surfaces for these systems.

For each orthogonal coordinate system the tables that follow specify

- coordinates,
- differentials of coordinates,
- transformations from curvilinear to Cartesian coordinates,
- unit vectors,
- vector components,
- metric coefficients leading to
- differential elements of length, and
- equations of coordinate surfaces.

From such tables all of the vector differential operators discussed in this guide may be applied to any of these orthogonal coordinate systems that one may choose. Following each table, perspective graphical views of selected orthogonal surfaces are shown for each specific coordinate system

The common four orthogonal systems—generalized orthogonal curvilinear coordinates (GOCCs), Cartesian (rectangular) coordinates, circular cylindrical coordinates, and spherical coordinates are presented Table B-1. The most general system of orthogonal coordinates is GOCCs shown in the first column of coordinates. Here all three metric coefficients are functions of all three coordinates. All orthogonal coordinate systems stem from GOCCs. Therefore, once any vector differential operator is determined in GOCCs it may readily be expressed in any orthogonal coordinate system by the proper substitution of coordinate variables and metric coefficients as specified by the tables.

At the other end of the generalization spectrum is Cartesian coordinates, shown in the second column of coordinates in Table B-1. The Cartesian system contains zero curvilinearity, that is, all three metric coefficients are unity and the DLs are just $dx$, $dy$, and $dz$ as noted in Sections B.1 and B.2 above. The first level of curvilinearity is circular cylindrical coordinates, having only one nonunity metric coefficient as explained in Section B.3. Spherical coordinates comprise the last column of the table, having two curvilinear coordinates. Figures B-4 through B-6 graphically present coordinate surfaces for these three coordinate systems, namely Cartesian, cylindrical, and spherical, respectively.

Table B-2 compares three other cylindrical systems—elliptic cylindrical, parabolic cylindrical, and bipolar cylindrical. Bipolar cylindrical is extremely useful in describing the fields associated with two-wire transmission lines. Coordinate surfaces for these three systems are shown in Figs. B-7 through B-9.

Table B-3 provides two spheroidal systems—prolate and oblate—as graphically depicted in Figs. B-10 and B-11. Figures B-10(a) and B-11(a) first show their respective spheroids alone. Figure B-10(b) displays two-sheeted hyperboloids that have the same foci as does the prolate spheroid. In contrast, Fig. B-11(b) displays the one-sheeted hyperboloid whose foci constitutes a concentric circle around the girth of the hyperboloid. This same circle is the foci of the oblate spheroid. Finally, the spheroid and hyperboloid surfaces are combined together with their respective $\phi$ planes, that are parallel with the $z$ axes, to form the orthogonal systems.

Tables B-4 and B-5 and Figs. B-12 and B-13 present coordinate parameters and surfaces, respectively, for sphero-conal and toroidal systems.

Although there are hundreds of orthogonal coordinate systems, this brief review is intended to provide selected examples that range from the common to the more esoteric.

**Table B-1** The Common Four Orthogonal Coordinate Systems

| Coordinate Systems→ & Parameters↓ | Generalized Curvilinear | Cartesian | Circular Cylindrical | Spherical |
|---|---|---|---|---|
| Coordinates and their range of values | $q_1$<br>$q_2$<br>$q_3$ | $-\infty < x < \infty$<br>$-\infty < y < \infty$<br>$-\infty < z < \infty$ | $0 \le r$ (or $r_c$) $< \infty$<br>$0 \le \phi \le 2\pi$<br>$-\infty < z < \infty$ | $0 \le r$ (or $r_s$) $< \infty$<br>$0 \le \theta \le \pi$<br>$0 \le \phi \le 2\pi$ |
| Transformation to Cartesian coordinates | $x = q_1$<br>$y = q_2$<br>$z = q_3$ | | $x = r\cos\phi$<br>$y = r\sin\phi$<br>$z = z$ | $x = r\sin\theta\cos\phi$<br>$y = r\sin\theta\sin\phi$<br>$z = r\cos\theta$ |
| Orthogonal Unit Vectors | $\hat{u}_1, \hat{u}_2, \hat{u}_3$ | $\hat{u}_x, \hat{u}_y, \hat{u}_z$ | $\hat{u}_r, \hat{u}_\phi, \hat{u}_z$ | $\hat{u}_r, \hat{u}_\theta, \hat{u}_\phi$ |
| Differentials of Coordinates | $dq_1, dq_2, dq_3$ | $dx, dy, dz$ | $dr, d\phi, dz$ or<br>$dr_c, d\phi, dz$ | $dr, d\theta, d\phi$ or<br>$dr_s, d\theta, d\phi$ |
| Components of the vector $\bar{A}$ | $A_1, A_2, A_3$ | $A_x, A_y, A_z$ | $A_r, A_\phi, A_z$ | $A_r, A_\theta, A_\phi$ |
| Metric, Lamé Coefficients, or scale factors | $h_1(q_1,q_2,q_3)$<br>$h_2(q_1,q_2,q_3)$<br>$h_3(q_1,q_2,q_3)$ | 1<br>1<br>1 | 1<br>$r$<br>1 | 1<br>$r$<br>$r\sin\theta$ |
| Differential Elements of Length | $dl_1 = h_1 dq_1$<br>$dl_2 = h_2 dq_2$<br>$dl_3 = h_3 dq_3$ | $dx$<br>$dy$<br>$dz$ | $dr$<br>$r\,d\phi$<br>$dz$ | $dr$<br>$r\,d\theta$<br>$r\sin\theta\,d\phi$ |
| Description of Coordinate Surfaces | | $x = x$<br>$y = y$<br>$z = z$<br>three orthogonal planes | $\sqrt{x^2+y^2} = r$<br>cylinders;<br>$\tan^{-1}\frac{y}{x} = \phi$<br>halfplanes from the z axis;<br>$z = z$, planes $\perp$ z axis. | $\sqrt{x^2+y^2+z^2} = r$<br>spheres;<br>$\cos^{-1}\frac{z}{r} = \theta$<br>z-axis cones<br>$\tan^{-1}\frac{y}{x} = \phi$<br>halfplanes from the z axis. |
| Coordinate Surface Graphics | | Fig. B-4 | Fig. B-5 | Fig. B-6 |

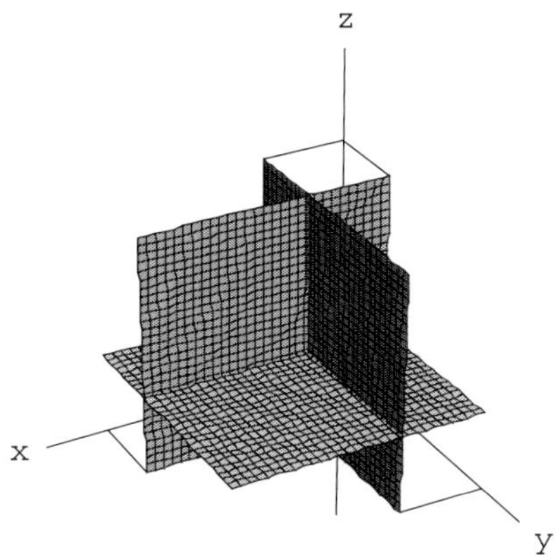


**Figure B-4**
Cartesian coordinate surfaces

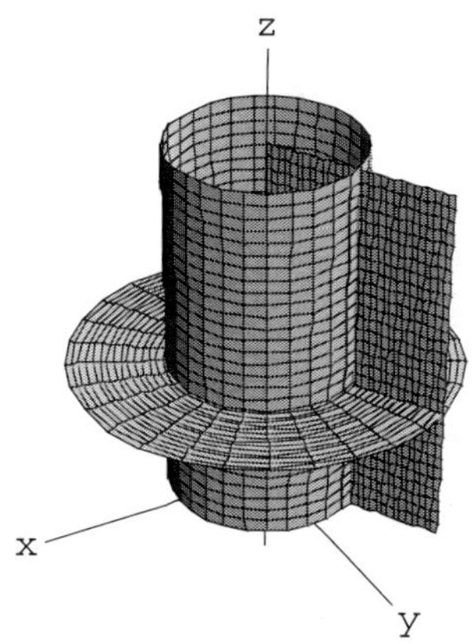


**Figure B-5**
Cylindrical coordinate surfaces

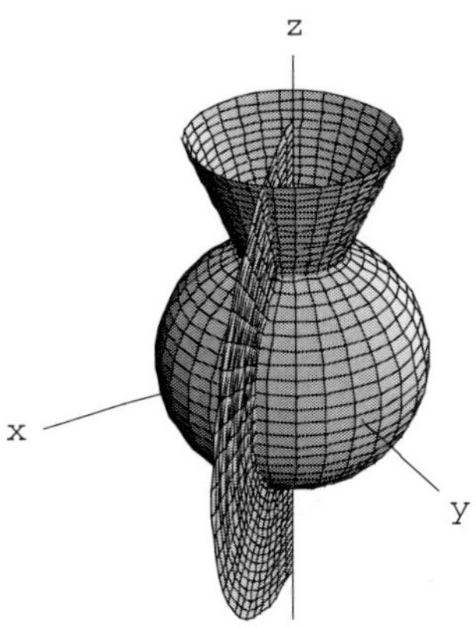


**Figure B-6**
Spherical coordinate surfaces

**Table B-2** Other Cylindrically based Coordinate Systems

| Coordinate Systems→ & Parameters↓ | Elliptic cylindrical | Parabolic Cylindrical | Bipolar cylindrical |
|---|---|---|---|
| Coordinates and their range of values | $\xi \geq 1$<br>$-1 \leq \eta \leq 1$<br>$-\infty \leq z \leq \infty$ | $-\infty \leq \xi \leq \infty$<br>$-\infty \leq \eta \leq \infty$<br>$-\infty \leq z \leq \infty$ | $-\infty \leq \xi \leq \infty$<br>$0 \leq \eta \leq 2\pi$<br>$-\infty \leq z \leq \infty$ |
| Transformation to Cartesian coordinates | $x = c_e \xi\eta$<br>$y = c_e\sqrt{(\xi^2-1)(1-\eta^2)}$<br>$z = z$ | $x = \xi\eta$<br>$y = \frac{1}{2}(\xi^2 - \eta^2)$<br>$z = z$ | $x = \frac{c_b \sinh \xi}{\cosh \xi - \cos \eta}$<br>$y = \frac{c_b \sin \eta}{\cosh \xi - \cos \eta}$<br>$z = z$ |
| Orthogonal Unit Vectors | $\hat{u}_\xi, \hat{u}_\eta, \hat{u}_z$ | $\hat{u}_\xi, \hat{u}_\eta, \hat{u}_z$ | $\hat{u}_\xi, \hat{u}_\eta, \hat{u}_z$ |
| Differentials of Coordinates | $d\xi, d\eta, dz$ | $d\xi, d\eta, dz$ | $d\xi, d\eta, dz$ |
| Components of the vector $\overline{A}$ | $A_\xi, A_\eta, A_z$ | $A_\xi, A_\eta, A_z$ | $A_\xi, A_\eta, A_z$ |
| Metric Coefficients, Lamé Coefficients, or scale factors | $c_e\sqrt{\frac{\xi^2-\eta^2}{\xi^2-1}}$<br>$c_e\sqrt{\frac{\xi^2-\eta^2}{1-\eta^2}}$<br>$1$ | $\sqrt{\xi^2+\eta^2}$<br>$\sqrt{\xi^2+\eta^2}$<br>$1$ | $\frac{c_b}{\cosh \xi - \cos \eta}$<br>$\frac{c_b}{\cosh \xi - \cos \eta}$<br>$1$ |
| Differential Elements of Length | $c_e\sqrt{\frac{\xi^2-\eta^2}{\xi^2-1}}\, d\xi$<br>$c_e\sqrt{\frac{\xi^2-\eta^2}{1-\eta^2}}\, d\eta$<br>$dz$ | $\sqrt{\xi^2+\eta^2}\, d\xi$<br>$\sqrt{\xi^2+\eta^2}\, d\eta$<br>$dz$ | $\frac{c_b}{\cosh \xi - \cos \eta} d\xi$<br>$\frac{c_b}{\cosh \xi - \cos \eta} d\eta$<br>$dz$ |
| Description of Coordinate Surfaces | $\frac{x^2}{\xi^2} + \frac{y^2}{\xi^2-1} = a^2$<br>elliptic cylinders;<br>$\frac{x^2}{\eta^2} - \frac{y^2}{1-\eta^2} = a^2$<br>hyperbolic cylinders;<br>$z = z$<br>planes ⊥ z axis. | $\xi^2 = \sqrt{x^2+y^2} - y$<br>hyperbolic cylinders;<br>$\eta^2 = \sqrt{x^2+y^2} + y$<br>hyperbolic cylinders;<br>$z = z$<br>planes ⊥ z axis. | $(x - c_b \coth\xi)^2 + y^2$<br>$= c_b \operatorname{csch}\xi^2$<br>x axis circular cylinders;<br>$x^2 + (y - c_b \cot\eta)^2$<br>$= c_b \csc\eta^2$<br>y axis circular cylinders;<br>$z = z$<br>planes ⊥ z axis. |
| Coordinate Surface Graphics | Fig. B-7 | Fig. B-8 | Fig B-9 |

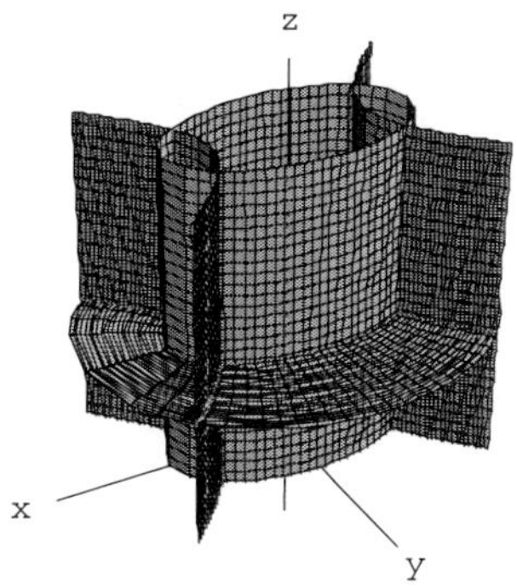


**Figure B-7**
Elliptic cylindrical surfaces

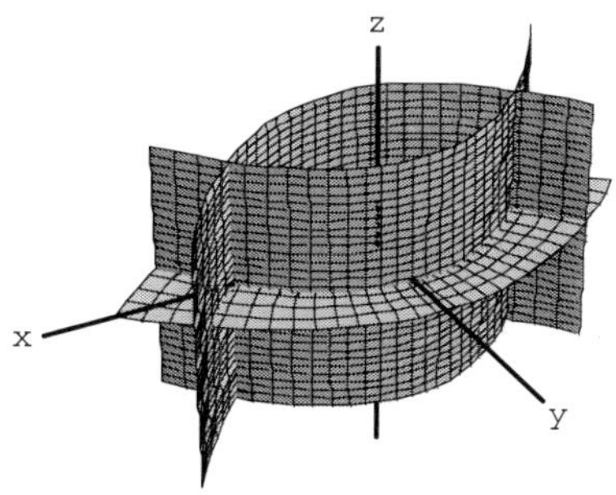


**Figure B-8**
Parabolic cylindrical surfaces

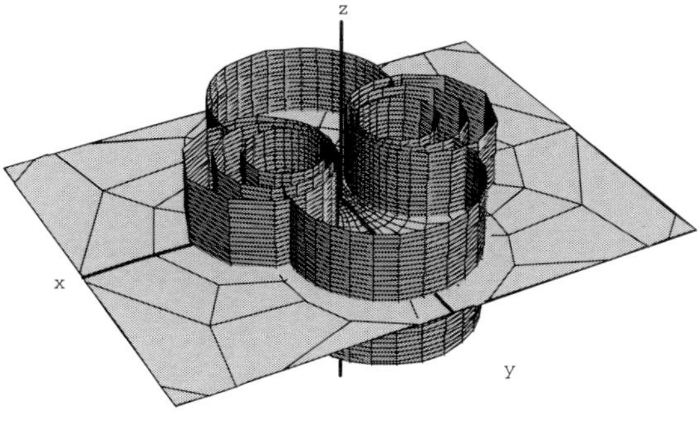


**Figure B-9**
Bipolar cylindrical surfaces

**Table B-3** Spheroidal Coordinate Systems

| Coordinate Systems→ & Parameters↓ | Confocal prolate spheroidal | Confocal oblate spheroidal |
|---|---|---|
| Coordinates and their range of values | $\xi \geq 1$<br>$0 \leq \eta \leq \pi$<br>$0 \leq \phi \leq 2\pi$ | $\xi \geq 1$<br>$0 \leq \eta \leq \pi$<br>$0 \leq \phi \leq 2\pi$ |
| Transformation to Cartesian coordinates | $x = c_p \sinh\xi \sin\eta \cos\phi$<br>$y = c_p \sinh\xi \sin\eta \sin\phi$<br>$z = c_p \cosh\xi \cos\eta$ | $x = c_o \cosh\xi \cos\eta \cos\phi$<br>$y = c_o \cosh\xi \cos\eta \sin\phi$<br>$z = c_o \sinh\xi \sin\eta$ |
| Orthogonal Unit Vectors | $\hat{u}_\xi, \hat{u}_\eta, \hat{u}_\phi$ | $\hat{u}_\xi, \hat{u}_\eta, \hat{u}_\phi$ |
| Differentials of Coordinates | $d\xi, d\eta, d\phi$ | $d\xi, d\eta, d\phi$ |
| Components of the vector $\bar{A}$ | $A_\xi, A_\eta . A_\phi$ | $A_\xi, A_\eta . A_\phi$ |
| Metric Coefficients, Lamé Coefficients, or scale factors | $c_p\sqrt{\sinh^2\xi - \sin^2\eta}$<br>$c_p\sqrt{\sinh^2\xi - \sin^2\eta}$<br>$c_p \sinh\xi \sin\eta$ | $c_o\sqrt{\sinh^2\xi - \sin^2\eta}$<br>$c_o\sqrt{\sinh^2\xi - \sin^2\eta}$<br>$c_o \cosh\xi \cos\eta$ |
| Differential Elements of Length | $c_p\sqrt{\sinh^2\xi - \sin^2\eta}\, d\xi$<br>$c_p\sqrt{\sinh^2\xi - \sin^2\eta}\, d\eta$<br>$c_p \sinh\xi \sin\eta\, d\phi$ | $c_o\sqrt{\sinh^2\xi - \sin^2\eta}\, d\xi$<br>$c_o\sqrt{\sinh^2\xi - \sin^2\eta}\, d\eta$<br>$c_o \cosh\xi \cos\eta\, d\phi$ |
| Description of Coordinate Surfaces | $\frac{x^2+y^2}{\sinh^2\xi} + \frac{z^2}{\cosh^2\xi} = c_p^{\ 2}$<br>prolate ellipsoids;<br>$\frac{x^2+y^2}{\sin^2\eta} - \frac{z^2}{\cos^2\eta} = -c_p^{\ 2}$<br>2-sheet hyperboloids;<br>$\tan^{-1}\frac{y}{x} = \phi$<br>halfplanes from the z axis. | $\frac{x^2+y^2}{\sinh^2\xi} + \frac{z^2}{\cosh^2\xi} = c_o^{\ 2}$<br>oblate ellipsoids;<br>$\frac{x^2+y^2}{\sin^2\eta} - \frac{z^2}{\cos^2\eta} = -c_o^{\ 2}$<br>1-sheet hyperboloids;<br>$\tan^{-1}\frac{y}{x} = \phi$<br>halfplanes from the z axis. |
| Coordinate Surface Graphics | Fig. B-10 | Fig. B-11 |

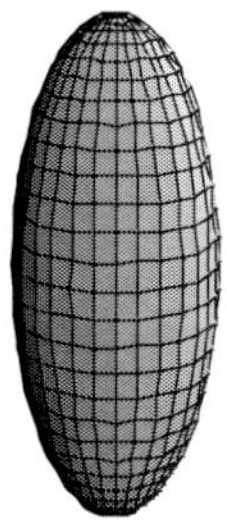

(a) Prolate spheroid

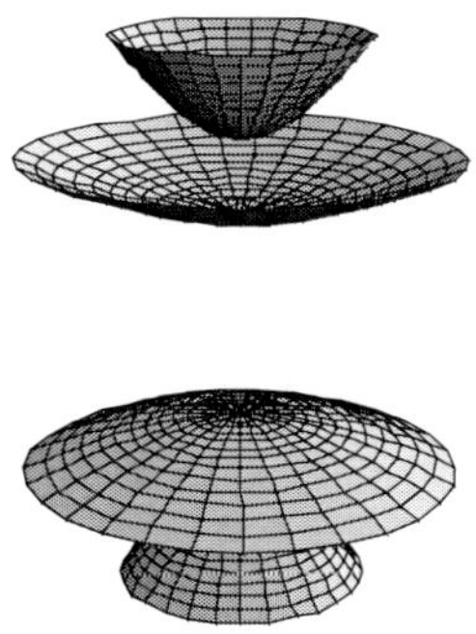

(b) Two-sheet hyperboloids

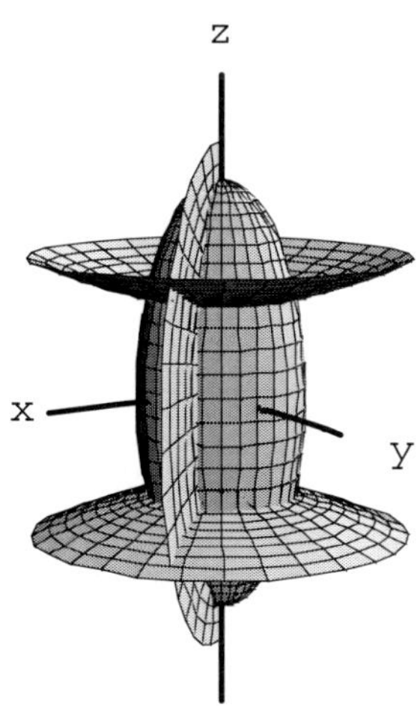


**Figure B-10**
Prolate spheroidal surfaces

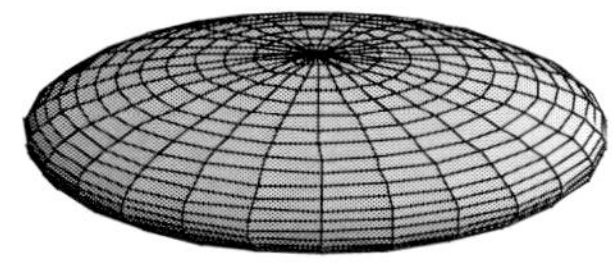

(a) Oblate spheroid

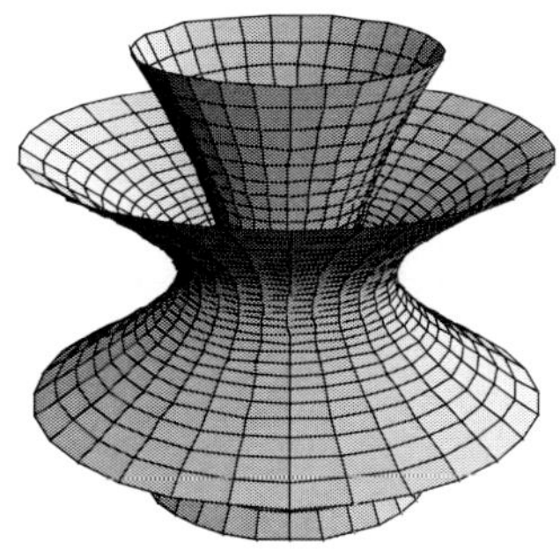

(b) One-sheet hyperboloids

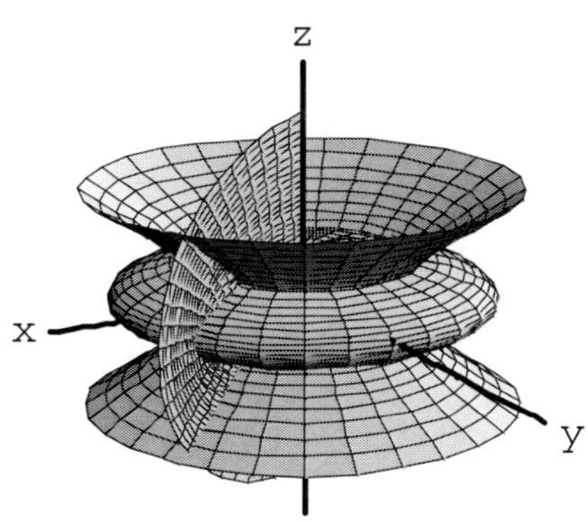


**Figure B-11**
Oblate spheroidal surfaces

**Table B-4** Sphero-conal Coordinate System

| Coordinate Systems→ & Parameters↓ | Sphero-conal |
|---|---|
| Coordinates and their range of values | $r \geq 0 \quad 0 \leq k \leq 1$<br>$0 \leq \theta \leq \pi \quad 0 \leq k' \leq 1$<br>$0 \leq \phi \leq 2\pi \quad k^2 + k'^2 = 1$ |
| Transformation to Cartesian coordinates | $x = r \sin\theta \cos\phi$<br>$y = r \sin\phi \sqrt{1 - k^2 \cos^2\theta}$<br>$z = r \cos\theta \sqrt{1 - k'^2 \sin^2\phi}$ |
| Orthogonal Unit Vectors | $\hat{u}_r, \hat{u}_\theta, \hat{u}_\phi$ |
| Differentials of Coordinates | $dr, d\theta, d\phi$ |
| Components of the vector $\overline{A}$ | $A_r, A_\theta, A_\phi$ |
| Metric Coefficients, Lamé Coefficients, or scale factors | $1$<br>$r\sqrt{\dfrac{k^2 \sin^2\theta + k'^2 \cos^2\phi}{1 - k^2 \cos^2\theta}}$<br>$r\sqrt{\dfrac{k^2 \sin^2\theta + k'^2 \cos^2\phi}{1 - k'^2 \sin^2\phi}}$ |
| Differential Elements of Length | $dr$<br>$r\sqrt{\dfrac{k^2 \sin^2\theta + k'^2 \cos^2\phi}{1 - k^2 \cos^2\theta}}\, d\theta$<br>$r\sqrt{\dfrac{k^2 \sin^2\theta + k'^2 \cos^2\phi}{1 - k'^2 \sin^2\phi}}\, d\phi$ |
| Description of Coordinate Surfaces | $\sqrt{x^2 + y^2 + z^2} = r$ spheres with radius $r = r_0$;<br>$\dfrac{x^2}{\tau^2} + \dfrac{y^2}{\tau^2 - b^2} - \dfrac{z^2}{c^2 - \tau^2} = 0$ elliptic cones $\tau = \tau_0$;<br>$\dfrac{x^2}{\lambda^2} - \dfrac{y^2}{b^2 - \lambda^2} - \dfrac{z^2}{c^2 - \lambda^2} = 0$ elliptic cones $\lambda = \lambda_0$;<br>$c^2 > \tau^2 > b^2 > \lambda^2 > 0.$ |
| Coordinate Surface Graphics | Fig. 12 |

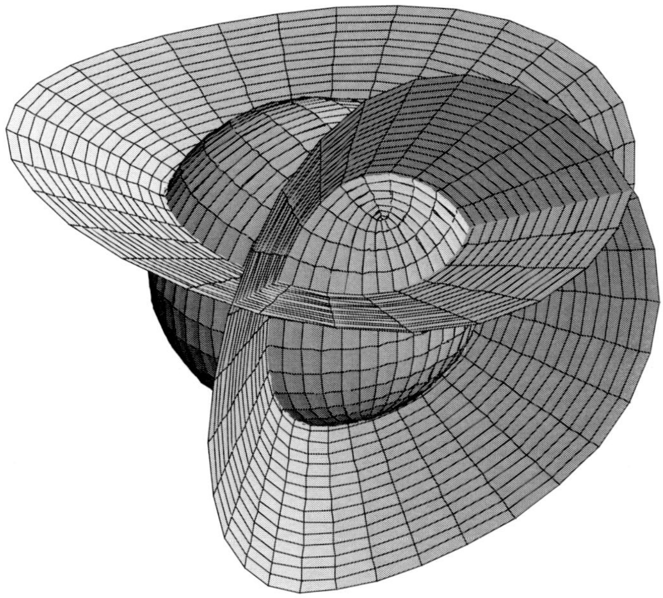

**Figure B-12** Sphero-conal coordinate surfaces

**Table B-5** Toroidal Coordinate System

| Coordinate Systems→ & Parameters↓ | Toroidal |
|---|---|
| Coordinates and their range of values | $-\infty \geq \xi \geq \infty$<br>$0 \leq \eta \leq 2\pi$<br>$0 \leq \phi \leq 2\pi$ |
| Transformation to Cartesian coordinates | $x = c_o \cosh\xi \cos\eta \cos\phi$<br>$y = c_o \cosh\xi \cos\eta \sin\phi$<br>$z = c_o \sinh\xi \sin\eta$ |
| Orthogonal Unit Vectors | $\hat{u}_\xi, \hat{u}_\eta, \hat{u}_\phi$ |
| Differentials of Coordinates | $d\xi, d\eta, d\phi$ |
| Components of the vector $\overline{A}$ | $A_\xi, A_\eta . A_\phi$ |
| Metric Coefficients, Lamé Coefficients, or scale factors | $c_o\sqrt{\sinh^2\xi - \sin^2\eta}$<br>$c_o\sqrt{\sinh^2\xi - \sin^2\eta}$<br>$c_o \cosh\xi \cos\eta$ |
| Differential Elements of Length | $c_o\sqrt{\sinh^2\xi - \sin^2\eta}\; d\xi$<br>$c_o\sqrt{\sinh^2\xi - \sin^2\eta}\; d\eta$<br>$c_o \cosh\xi \cos\eta\; d\phi$ |
| Description of Coordinate Surfaces | $\sin^2\eta\left[\left(x^2+y^2\right)-\left(z - a\cot\eta\right)^2\right]^2 = c_t^{\,2}$<br>spheres;<br>$\sinh^2\xi\left[\sqrt{x^2+y^2} - a\coth\xi\right]^2 + z^2 = c_t^{\,2}$<br>tores or anchor rings;<br>$\tan^{-1}\frac{y}{x} = \phi$<br>halfplanes from the z axis. |
| Coordinate Surface Graphics | Fig. 13 |

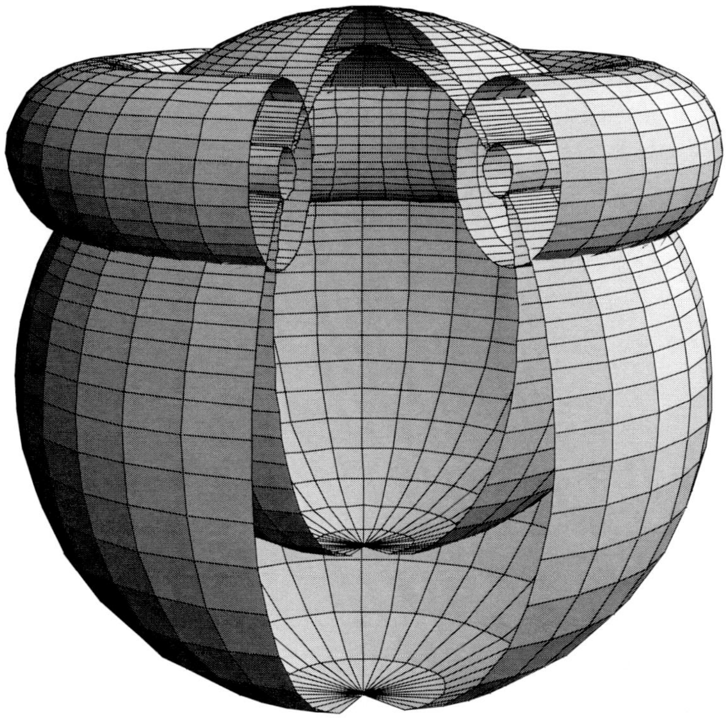

**Figure B-13** Toroidal coordinate surfaces

## References

1. Granino A. Korn and Theresa M. Korn, *Mathematical Handbook for Scientists and* Engineers, 2nd ed., McGraw-Hill, New York (1968).
2. Parry Moon and Domina Eberlie Spencer, *Field Theory for Engineers*, D. Van Nostrand Co., Inc., New York (1961).
3. E. W. Hobson, *The Theory of Spherical and Ellipsoidal Harmonics*, Chelsea Pub. Co., New York (1955).
4. J. H. Heinbockel, *Introduction to Tensor Calculus and Continuum Mechanics*, Self-published through Trafford Pub. (2001).
5. Julius Adams Stratton, *Electromagnetic Theory*, McGraw-Hill, New York (1941).
6. Philip M. Morse and Herman Feshbach, *Methods of Theoretical Physics*, Vols. I and II (1953).
7. Parry Moon and Domina Eberlie Spencer, *Field Theory Handbook*, Springer-Verlag, Berlin (1961).
8. A. N. Tikhonov, A. A. Samarskii, *Uravneniya Matematicheskoy Phiziki* (Moscow: Nauka Publisher, 1966).
9. V. I. Blokh, *Teoriya Uprugosti* (Ukraine: Harkov Unibersitet Izdatelitvo, 1964).

## Appendix C

# Intermediate Tensor Calculus in Support of Chapters 3 and 4

Andrey Beyle* and Bernard Maxum

This appendix details the proofs of several important postulations used in Chapters 3 and 4. Each of these proofs is performed for tensors of general rank expressed in explicit standard notation.** Section C.1 provides the precise formulation of the use of this notation for general rank tensors. In addition the dot-, cross-, and tensor-product operations are shown for general rank tensors with the introduction of a generalized operator symbol denoted by an asterisk "*" used to denote any one of these three operators.

Section C.2 deals with properties of vector differential operators of first and second order. It has six subsections. In Section C.2.1 the rank progression, cited at the bottom of page 3-21 for dot-, cross-, and tensor-product operators, is demonstrated for vectors and general-rank tensors. The subsequent four subsections provide proofs pertaining to crucial properties of first- and second-order vector differential operations on general rank tensor operands used throughout Chapter 4, including Tables 4-1 and 4-2. Finally, Section C.2.6 gives a proof of the Lagrange identity [Eq. (4.7-25)] for tensors of general rank.

The divergence operator of Eq. (4.7-7) in Section 4.7.4(a) was developed inductively. It was applied to a vector [Eq. (4.4-22)] and a dyadic [Eq. (4.7-9)] and cited for application to any tensor. Section C.3 gives a deductive proof of its general application to a tensor.

Finally, in Section C.4 we demonstrate that for all but Cartesian coordinates it is necessary to consider the derivative nature of the nabla operator as well as its dot-, cross-, and tensor-product nature. Thus, the tendency of depicting the divergence, curl, and gradient operators as analogous to dot, cross, and tensor products is shown to be false (except for Cartesian coordinates).

---

* Co-author for this appendix

** See Section 1.1 for a description of various notational forms. Explicit standard notation is described in Section 1.1.7(a) on page 1-11.

## C.1 Explicit Standard Notation for General Rank Tensors

We first represent a tensor of general rank in explicit standard notation as

$$\overset{\cdots}{\overline{\overline{\overline{T}}}} = \sum_{i=1}^{3}\sum_{j=1}^{3}\sum_{k=1}^{3}\cdots \left(T_{ijk\cdots}\,\hat{u}_i\hat{u}_j\hat{u}_k\cdots\right) \tag{C.1-1}$$

The general rank tensor $\overset{\cdots}{\overline{\overline{\overline{T}}}}$ is shown with three overbars and the ellipsis "$\cdots$" overmarking indicating a general number of overbars corresponding to the tensor rank. Since three overbars are explicitly depicted, three summations are shown followed by an ellipsis indicating the number of summations corresponding to the rank. The summations are taken from 1 to 3 for 3D space. The symbol $T_{ijk\cdots}$ represents the $ijk\cdots^{\text{th}}$ scalar component of the tensor $\overset{\cdots}{\overline{\overline{\overline{T}}}}$. The "$\hat{u}_i\hat{u}_j\hat{u}_k\cdots$" represents the generalized unitary tensor and are shown as "direct" products.* Another form of Eq. (C.1-1) can show the "tensor" products explicitly. Thus (C.1-1) is equivalently expressed as

$$\overset{\cdots}{\overline{\overline{\overline{T}}}} = \sum_{i=1}^{3}\sum_{j=1}^{3}\sum_{k=1}^{3}\cdots \left(T_{ijk\cdots}\,\hat{u}_i\otimes\hat{u}_j\otimes\hat{u}_k\otimes\cdots\right) \tag{C.1-2}$$

Using the so-called Rule of Einstein,[1] the summations over repeated indices are implied and are omitted for simplicity. Therefore Eq. (C.1-2) could be written as

$$\overset{\cdots}{\overline{\overline{\overline{T}}}} = T_{ijk\cdots}\,\hat{u}_i\otimes\hat{u}_j\otimes\hat{u}_k\otimes\cdots \tag{C.1-3}$$

Equation (C.1-1), when combined with the Rule of Einstein, yields an even more concise form to express our general rank tensor $\overset{\cdots}{\overline{\overline{\overline{T}}}}$ as

$$\overset{\cdots}{\overline{\overline{\overline{T}}}} = T_{ijk\cdots}\,\hat{u}_i\hat{u}_j\hat{u}_k\cdots \tag{C.1-4}$$

In order to operate on general rank tensors, both the beginning and the end indices need to be explicit. Thus a $z$-rank tensor $\overset{\cdots}{\overline{\overline{\overline{T}}}}$ operating on an $\omega$-rank tensor $\overset{\cdots}{\overline{\overline{\overline{U}}}}$ could be expressed as

---

* See the footnote on page 2-7 from Section 2.4.3 describing the "direct" product as synonymous with "tensor" product, denoted with the symbol $\otimes$, which implies that all possible combinations of components are taken.

$$\begin{aligned}&\left(T_{ijk\cdots xyz}\hat{u}_i\hat{u}_j\hat{u}_k\cdots\hat{u}_x\hat{u}_y\hat{u}_z\right)*\left(U_{\alpha\beta\gamma\cdots\chi\psi\omega}\hat{u}_\alpha\hat{u}_\beta\hat{u}_\gamma\cdots\hat{u}_\chi\hat{u}_\psi\hat{u}_\omega\right)\\&=\left(T_{ijk\cdots xyz}U_{\alpha\beta\gamma\cdots\chi\psi\omega}\hat{u}_i\hat{u}_j\hat{u}_k\cdots\hat{u}_x\hat{u}_y\right)\left(\hat{u}_z*\hat{u}_\alpha\right)\hat{u}_\beta\hat{u}_\gamma\cdots\hat{u}_\chi\hat{u}_\psi\hat{u}_\omega\end{aligned} \qquad \text{(C.1-5)}$$

where the asterisk "$*$" denotes either a dot-, cross-, or tensor-product operation. In other words, the last unit vector of the first tensor operates on the first unit vector of the second tensor.

*Application to the dyadic-dyadic dot product*

As an example, let us first apply Eq. (C.1-5) to the case of the dyadic dot product with another dyadic that was covered in Section 3.4.2. In this case Eq. (C.1-5) is written as

$$\left(T_{ij}\hat{u}_i\hat{u}_j\right)\cdot\left(U_{mn}\hat{u}_m\hat{u}_n\right)=T_{ij}U_{mn}\delta_{jm}\hat{u}_i\hat{u}_n=T_{ij}U_{jn}\hat{u}_i\hat{u}_n \qquad \text{(C.1-6)}$$

where the Kronecker delta $\delta_{jm}$ was defined by Eq. (2.4-12). This expression is equivalent to Eq. (3.4-12), which is apparent after substituting $T$ for $A$, $U$ for $B$ and adjusting for the change to tensor notation. That is, the tensor notation representation for Eq. (C.1-6) is

$$T_{ij}\cdot U_{jn}=V_{in} \qquad \text{(C.1-7)}$$

where the three double-subscripted quantities are tensors of rank two, i.e. dyadics, and where $V_{in}$ in tensor notation is $T_{ij}U_{jn}\hat{u}_i\hat{u}_n$ in explicit standard notation. Note also that the $T_{i\,j}$ in Eq. (C.1-6) is a scalar component of the dyadic tensor $\overline{\overline{T}}$ whereas $T_{ij}$ in Eq. (C.1-7) ***is*** the dyadic tensor $\overline{\overline{T}}$. As explained in the paragraph following Eq. (3.3-5), these are distinguishable by the spacing of the subscripts.

If we repeat the above exercise by first applying the form (C.1-1), the manipulation of the summation signs becomes explicit. Thus the dot product takes the form

$$\begin{aligned}\bar{\bar{T}}\cdot\bar{\bar{U}} &= \left[\sum_{i=1}^{3}\sum_{j=1}^{3}\left(T_{ij}\hat{u}_i\hat{u}_j\right)\right]\cdot\left[\sum_{m=1}^{3}\sum_{n=1}^{3}\left(U_{mn}\hat{u}_m\hat{u}_n\right)\right] \\ &= \sum_{i=1}^{3}\sum_{j=1}^{3}\sum_{m=1}^{3}\sum_{n=1}^{3}T_{ij}U_{mn}\delta_{jm}\hat{u}_i\hat{u}_n = \sum_{i=1}^{3}\sum_{j=1}^{3}\sum_{n=1}^{3}T_{ij}U_{jn}\hat{u}_i\hat{u}_n\end{aligned} \tag{C.1-8}$$

One can see how the dot product subtracts two from the sum of the ranks of the two dyadics. That is, $\bar{\bar{T}}$ and $\bar{\bar{U}}$ each have a rank of two depicted by the unit dyads $\hat{u}_i\hat{u}_j$ and $\hat{u}_m\hat{u}_n$, respectively, and of the resulting 81 components, 54 are zero due to the inside dot products $\hat{u}_j\cdot\hat{u}_m$ when $j\neq m$. The remaining 27 terms survive when $j=m$. The Kronecker delta $\delta_{jm}$, defined by Eq. (2.4-12), essentially says all of this in one convenient symbol. Of the surviving 27 unit dyads $\hat{u}_i\hat{u}_n$, only nine are unique, namely the nine combinations of $i$ and $n$ running from 1 to 3. Thus from the r.h.s. of Eq. (C.1-8), the resultant has a rank of two. Notice how the Kronecker delta "kills" the $m^{th}$ summation sign leaving the inner $j$ summation to be taken over both $T_{ij}$ and $U_{jn}$. Notice also that the r.h.s. denotes a dyadic with nine components arising from the $i$ and $n$ summations. Notice also that each of the nine components contains three scalar terms arising from the $j$ summation.

The above discussion may become more apparent for those new to tensors, by expanding Eq. (C.1-8) into its terms and components as

$$\begin{aligned}\bar{\bar{T}}\cdot\bar{\bar{U}} &= \begin{bmatrix} \hat{u}_1\hat{u}_1(T_{11}U_{11}+T_{12}U_{21}+T_{13}U_{31}) & \hat{u}_1\hat{u}_2(T_{11}U_{12}+T_{12}U_{22}+T_{13}U_{32}) & \hat{u}_1\hat{u}_3(T_{11}U_{13}+T_{12}U_{23}+T_{13}U_{33}) \\ \hat{u}_2\hat{u}_1(T_{21}U_{11}+T_{22}U_{21}+T_{23}U_{31}) & \hat{u}_2\hat{u}_2(T_{21}U_{12}+T_{22}U_{22}+T_{23}U_{32}) & \hat{u}_2\hat{u}_3(T_{21}U_{13}+T_{22}U_{23}+T_{23}U_{33}) \\ \hat{u}_3\hat{u}_1(T_{31}U_{11}+T_{32}U_{21}+T_{33}U_{31}) & \hat{u}_3\hat{u}_2(T_{31}U_{12}+T_{32}U_{22}+T_{33}U_{32}) & \hat{u}_3\hat{u}_3(T_{31}U_{13}+T_{32}U_{23}+T_{33}U_{33}) \end{bmatrix} \\ &= \begin{bmatrix} \hat{u}_1\hat{u}_1(V_{11}) & \hat{u}_1\hat{u}_2(V_{12}) & \hat{u}_1\hat{u}_3(V_{13}) \\ \hat{u}_2\hat{u}_1(V_{21}) & \hat{u}_2\hat{u}_2(V_{22}) & \hat{u}_2\hat{u}_3(V_{23}) \\ \hat{u}_3\hat{u}_1(V_{31}) & \hat{u}_3\hat{u}_2(V_{32}) & \hat{u}_3\hat{u}_3(V_{33}) \end{bmatrix}\end{aligned} \tag{C.1-9}$$

where the scalar components of $\bar{\bar{V}}$ are related to the scalar components of $\bar{\bar{T}}$ and $\bar{\bar{U}}$ through Eq. (C.1-7) as

$$V_{ik} = \sum_{j=1}^{3} T_{ij}U_{jk} \tag{C.1-10}$$

## C.2 Properties of First- and Second-Order Vector Differential Operators on Tensors

This section demonstrates key properties of first- and second-order vector differential operations on scalars, vectors, dyadics, and general-rank tensors cited in Tables 4-1 and 4-2 and used throughout Chapter 4. In the first subsection (Section C.2.1) the rank progression cited in Table 4-1 resulting from div, curl, and grad is shown. The zero values postulated in the table of second-order differential operator resultant forms given in Table 4-2 are developed in the next two subsections. That the divergence of the curl is zero (**DC**$T = 0$) and the curl of the gradient is zero (**CG**$T = 0$) are shown in Sections C.2.2 and C.2.3, respectively. The reverse of these operations, however, are, in general, nonzero, namely (**CD**$T \neq 0$) and (**GC**$T \neq 0$). These are demonstrated in Sections C.2.4 and C.2.5, respectively. Finally, the application of the Lagrange identity to general-rank tensors is demonstrated in Section C.2.6.

These demonstrations and proofs are made for the general-rank tensor. However, even though the operand is generalized, the coordinate system need not be, for if we can show these proofs in one orthogonal coordinate system it must be so for all orthogonal systems. We choose to do these proofs in Cartesian coordinates to avoid the cumbersomeness of having to deal with coordinate derivatives of unit vectors. In doing so, there is no loss in generality for the purpose of these proofs.

### C.2.1 First-order vector differential operators with vector and generalized tensor operands

In Cartesian coordinates the nabla operator can be written using Einstein Rule[1]

$$\nabla = \hat{u}_i \nabla_i = \hat{u}_i \frac{\partial}{\partial x_i} \tag{C.2-1}$$

and follows the same rules as a vector. Therefore the divergence and curl of the vector $\overline{A}$ become

$$\nabla \cdot \overline{A} = \left( \hat{u}_i \frac{\partial}{\partial x_i} \right) \cdot A_k \hat{u}_k = \frac{\partial A_k}{\partial x_i} \left( \hat{u}_i \cdot \hat{u}_k \right) = \frac{\partial A_k}{\partial x_i} \delta_{ik} = \frac{\partial A_i}{\partial x_i} \tag{C.2-2}$$

$$\nabla \times \bar{A} = \left( \hat{u}_i \frac{\partial}{\partial x_i} \right) \times A_k \hat{u}_k = \frac{\partial A_k}{\partial x_i} \left( \hat{u}_i \times \hat{u}_k \right) = \frac{\partial A_k}{\partial x_i} \in_{ikl} \hat{u}_l \qquad \text{(C.2-3)}$$

where the Levi-Civita symbol $\in_{ikl}$, which was defined in Eq. (2.4-27), is critical to the "cross" operation used in cross products and curl operators. Finally, the gradient of the vector $\bar{A}$ becomes

$$\begin{aligned} \nabla \otimes \bar{A} &= \left( \hat{u}_i \frac{\partial}{\partial x_i} \right) \otimes A_k \hat{u}_k = \frac{\partial A_k}{\partial x_i} \left( \hat{u}_i \otimes \hat{u}_k \right) \\ &= \nabla \bar{A} = \frac{\partial A_k}{\partial x_i} \left( \hat{u}_i \hat{u}_k \right) \end{aligned} \qquad \text{(C.2-4)}$$

where the second line is depicted in "direct" product form (without the tensor-product symbol). In the above three expressions the Cartesian coordinate derivatives of Cartesian unit vectors are all zero and thus are omitted. This is the advantage of performing this development in Cartesian coordinates.*

Next the divergence, curl, and gradient vector operators are developed for general rank tensors. Expressions similar to Eq. (C.2-2) through (C.2-4) can be obtained for the results of application of the nabla operator to general rank tensors because only the first unit vector of the tensor will be participating in the operation. Using the form (C.1-4) to express the generalized tensor $\overset{\vdots}{\bar{\bar{\bar{B}}}}$ and replacing the first tensor with the vector nabla operator (C.2-1), we may apply Eq. (C.1-5) to obtain the divergence, curl, and gradient operations on a generalized tensor. Thus, the divergence of a generalized tensor in Cartesian coordinates is

$$\begin{aligned} \nabla \cdot \overset{\vdots}{\bar{\bar{\bar{B}}}} &= \left( \hat{u}_i \frac{\partial}{\partial x_i} \right) \cdot B_{jkl\cdots} \hat{u}_j \hat{u}_k \hat{u}_l \cdots \\ &= \frac{\partial B_{jkl\cdots}}{\partial x_i} \left( \hat{u}_i \cdot \hat{u}_j \right) \hat{u}_k \hat{u}_l \cdots \\ &= \frac{\partial B_{jkl\cdots}}{\partial x_i} \delta_{ij} \hat{u}_k \hat{u}_l \cdots = \frac{\partial B_{ikl\cdots}}{\partial x_i} \hat{u}_k \hat{u}_l \cdots \end{aligned} \qquad \text{(C.2-5)}$$

---

* Caution! This procedure is valid only when Cartesian coordinates are being used because only in Cartesian coordinates are all coordinate derivatives of unit vectors zero. For any curvilinear coordinate system, even cylindrical coordinates, coordinate derivatives of unit vectors must be considered.

Likewise, the curl of a generalized tensor in Cartesian coordinates is

$$\nabla\times\overset{\vdots}{\overline{\overline{\overline{B}}}}=\left(\hat{u}_i\frac{\partial}{\partial x_i}\right)\times B_{jkl\cdots}\hat{u}_j\hat{u}_k\hat{u}_l\cdots$$
$$=\frac{\partial B_{jkl\cdots}}{\partial x_i}\left(\hat{u}_i\times\hat{u}_j\right)\hat{u}_k\hat{u}_l\cdots=\frac{\partial B_{jkl\cdots}}{\partial x_i}\in_{ijh}\hat{u}_h\hat{u}_k\hat{u}_l\cdots \tag{C.2-6}$$

and the gradient of a generalized tensor in Cartesian coordinates is

$$\nabla\overset{\vdots}{\overline{\overline{\overline{B}}}}=\left(\hat{u}_i\frac{\partial}{\partial x_i}\right)B_{jkl\cdots}\hat{u}_j\hat{u}_k\hat{u}_l\cdots$$
$$=\frac{\partial B_{jkl\cdots}}{\partial x_i}\left(\hat{u}_i\hat{u}_j\right)\hat{u}_k\hat{u}_l\cdots=\frac{\partial B_{jkl\cdots}}{\partial x_i}\hat{u}_i\hat{u}_j\hat{u}_k\hat{u}_l\cdots \tag{C.2-7}$$

By taking careful notice of the final forms of each of the three vector differential operators above, it is readily determined that

- The divergence decreases the rank of the operand tensor by one,
- The curl conserves the rank of the operand tensor, and
- The gradient increases the rank of the operand tensor by one.

Thus, the right-hand column of Table 4-1 is demonstrated (as are the preceding columns, which are special cases of the right-hand column).

### C.2.2 Proof that the divergence of the curl of any tensor is zero

By performing proofs of the zeros that appear in Table 4-2 for **DC** and **CG** operations on the generalized tensor operand, we not only demonstrate the zeros of part (d) of that table, but also the zeros in parts (c) and (b) and the zero in (a) as well. This is because the operands for these earlier parts are special cases of tensors. In this subsection we show that the divergence of the curl of any tensor is zero. This applies to tensors having a rank $n_R \geq 1$, which excludes scalars, since the curl of a scalar is nonexistent. See Table 4-2(a).

Let's replace the dummy index $i$ in Eq. (C.2-6), the expression for the curl of a general-rank tensor, by another symbol "$g$." This is necessary because we will reserve the index $i$ for the second operation, namely the divergence. The result of the curl operation can be denoted as a new tensor $\overset{\vdots}{\overline{\overline{\overline{C}}}}$. Thus from Eq. (C.2-6),

$$\overset{\vdots}{\overline{\overline{\overline{C}}}}=\nabla\times\overset{\vdots}{\overline{\overline{\overline{B}}}}=\frac{\partial B_{jkl\cdots}}{\partial x_g}\in_{gjh}\hat{u}_h\hat{u}_k\hat{u}_l\cdots \tag{C.2-8}$$

which, of course, is a different tensor but having the same rank as the operand tensor $\overset{\vdots}{\overline{\overline{\overline{B}}}}$.

Next, we express the divergence of $\overset{\vdots}{\overline{\overline{\overline{C}}}}$, using Eq. (C.2-5) as

$$\begin{aligned}\nabla\cdot\overset{\vdots}{\overline{\overline{\overline{C}}}}&=\nabla\cdot\left(\nabla\times\overset{\vdots}{\overline{\overline{\overline{B}}}}\right)=\nabla\cdot\left(\frac{\partial B_{jkl\cdots}}{\partial x_g}\in_{gjh}\hat{u}_h\hat{u}_k\hat{u}_l\cdots\right)\\&=\frac{\partial^2 B_{jkl\cdots}}{\partial x_g\partial x_i}\left(\hat{u}_i\cdot\hat{u}_h\right)\in_{gjh}\hat{u}_k\hat{u}_l\cdots\\&=\frac{\partial^2 B_{jkl\cdots}}{\partial x_g\partial x_i}\delta_{ih}\in_{gjh}\hat{u}_k\hat{u}_l\cdots\\&=\nabla\cdot\left(\nabla\times\overset{\vdots}{\overline{\overline{\overline{B}}}}\right)=\frac{\partial^2 B_{jkl\ldots}}{\partial x_g\partial x_i}\in_{gji}\hat{u}_k\hat{u}_l\cdots\end{aligned} \tag{C.2-9}$$

Expanding Eq. (C.2-9) we have

$$\begin{aligned}&\in_{gji}\left(\frac{\partial^2}{\partial x_g\partial x_i}\right)\\&=-\left(\frac{\partial^2}{\partial x_1\partial x_2}\right)+\left(\frac{\partial^2}{\partial x_2\partial x_1}\right)+\left(\frac{\partial^2}{\partial x_1\partial x_3}\right)\\&\quad-\left(\frac{\partial^2}{\partial x_3\partial x_1}\right)-\left(\frac{\partial^2}{\partial x_2\partial x_3}\right)+\left(\frac{\partial^2}{\partial x_3\partial x_2}\right)=0\end{aligned} \tag{C.2-10}$$

Because of the mixed derivative theorem [Eq. (1.3-11), p. 1-20] each of the first-second, third-fourth and fifth-sixth paired terms cancel. Therefore

$$\mathbf{DC}\,\overset{\cdots}{\overline{\overline{\overline{B}}}} = \nabla \cdot \left( \nabla \times \overset{\cdots}{\overline{\overline{\overline{B}}}} \right) = 0 \qquad \text{(C.2-11)}$$

### C.2.3 Proof that the curl of the gradient of any tensor is zero

Somewhat in parallel with the procedure in the preceding proof, we next show that the curl of the gradient of a tensor vanishes. Since the first operation is the gradient of a generalized tensor $\overset{\cdots}{\overline{\overline{\overline{B}}}}$, let us define a new tensor $\overset{\cdots}{\overline{\overline{\overline{D}}}}$ from Eq. (C.2-7) as

$$\overset{\cdots}{\overline{\overline{\overline{D}}}} = \nabla \overset{\cdots}{\overline{\overline{\overline{B}}}} = \frac{\partial B_{jkl\cdots}}{\partial x_g} \hat{u}_g \hat{u}_j \hat{u}_k \hat{u}_l \cdots \qquad \text{(C.2-12)}$$

Then applying Eq. (C.2-6) we express the curl of $\overset{\cdots}{\overline{\overline{\overline{D}}}}$ as

$$\begin{aligned} \nabla \times \overset{\cdots}{\overline{\overline{\overline{D}}}} &= \nabla \times \left( \nabla \overset{\cdots}{\overline{\overline{\overline{B}}}} \right) \\ &= \frac{\partial^2 B_{jkl\cdots}}{\partial x_g \partial x_i} \left( \hat{u}_i \times \hat{u}_g \right) \hat{u}_j \hat{u}_k \cdots \\ &= \frac{\partial^2 B_{jkl\cdots}}{\partial x_g \partial x_i} \in_{igh} \hat{u}_h \hat{u}_j \hat{u}_k \cdots \end{aligned} \qquad \text{(C.2-13)}$$

Again, expanding Eq. (C.2-13) we have second-order derivatives

$$\in_{igh}\left(\frac{\partial^2}{\partial x_g \partial x_i}\right)$$

$$=+\left(\frac{\partial^2}{\partial x_1 \partial x_2}\right)-\left(\frac{\partial^2}{\partial x_2 \partial x_1}\right)-\left(\frac{\partial^2}{\partial x_1 \partial x_3}\right) \tag{C.2-14}$$

$$+\left(\frac{\partial^2}{\partial x_3 \partial x_1}\right)+\left(\frac{\partial^2}{\partial x_2 \partial x_3}\right)-\left(\frac{\partial^2}{\partial x_3 \partial x_2}\right)=0$$

and, again, because of the mixed derivative theorem all of the terms cancel. Therefore

$$\mathbf{CG}\,\overset{\vdots}{\overline{\overline{\overline{B}}}} = \nabla\times\left(\nabla\overset{\vdots}{\overline{\overline{\overline{B}}}}\right)=0 \tag{C.2-15}$$

This applies to tensors having a rank $n_R \geq 0$, which comprises of all tensors. This includes scalars (and vectors), since the gradient of a scalar is a vector and the curl of that vector may be taken.

### C.2.4 Demonstration that the curl of the divergence of any tensor is in general nonzero

In Section C.2.2 it was shown that the divergence of the curl of a general-rank tensor is always zero. The physical rationale for such a result might be argued in the following manner: Since the curl of a vector field behaves somewhat like a cross product, i.e., it is primarily a property mainly transverse to the vector field, and since the divergence of that resulting vector field is a measure of the net outward flux of that new field, it is not entirely surprising that that net flux would be zero. That argument can be extended to tensor fields with the exception that the "transverse" nature of the curl would be with respect to the directivity of the first unit vector of the multiply directed tensor.

This section deals with the reverse operation — the curl of the divergence. Although similar arguments may be made with respect to the transverse nature of the curl, we are dealing here with the second directional unit vector of the tensor, not the first. This is due to the fact that the first directional unit vector is annihilated by the dot-product nature of the divergence operation. Since this second-level directionality is arbitrary, there is no reason to expect that taking the curl after the divergence would have a vanishing result. Further, the curl of the divergence of a vector is nonexistent because the divergence yields a scalar, and one cannot take the curl of a scalar. Therefore Table 4-2(b) is marked accordingly. [The operation on a scalar is, of course, nonexistent because the divergence operation is nonexistent in the first place. See Table 4-2(a)].

The above argument becomes apparent mathematically. We first construct a tensor $\overset{\cdots}{\overline{\overline{\overline{E}}}}$ defined as the divergence of our operand tensor $\overset{\cdots}{\overline{\overline{\overline{B}}}}$. We will use Eq. (C.2-5) for the first operation; however, with a view toward using Eq. (C.2-6) for the second operation, we will change the index $i$ in Eq. (C.2-5) to $g$ as before.

$$\begin{aligned} &\overset{\cdots}{\overline{\overline{\overline{E}}}} = \nabla \cdot \overset{\cdots}{\overline{\overline{\overline{B}}}} = \\ &\frac{\partial B_{jkl\cdots}}{\partial x_g} \delta_{gj} \hat{u}_k \hat{u}_l \cdots \\ &= \frac{\partial B_{gkl\cdots}}{\partial x_g} \hat{u}_k \hat{u}_l \cdots \end{aligned} \tag{C.2-16}$$

where $\overset{\cdots}{\overline{\overline{\overline{B}}}}$ is a tensor of the rank $n_R \geq 2$ for reasons given in the second paragraph of this section. After substitution into Eq. (C.2-6) for the curl of the tensor of rank $n_R \geq 1$, we have

$$\begin{aligned} &\nabla \times \overset{\cdots}{\overline{\overline{\overline{E}}}} = \nabla \times \left( \nabla \cdot \overset{\cdots}{\overline{\overline{\overline{B}}}} \right) = \\ &\frac{\partial^2 B_{jkl\cdots}}{\partial x_g \partial x_i} \delta_{gj} \left( \hat{u}_i \times \hat{u}_k \right) \hat{u}_l \hat{u}_m \cdots = \\ &\frac{\partial^2 B_{jkl\cdots}}{\partial x_g \partial x_i} \delta_{gj} \in_{ikh} \hat{u}_h \hat{u}_l \hat{u}_m \cdots \end{aligned} \tag{C.2-17}$$

Since our objective is to show that the *curl div* is in general not zero, we need only show that one term is not zero. To do this we arbitrarily pick a combination of indices. If we examine the $\partial x_1 \partial x_2$ and sum over $k$, then we let $g=1$, $j=1$, $i=2$ and Eq. (C.2-17) becomes

$$\frac{\partial^2 B_{jkl\cdots}}{\partial x_1 \partial x_2}\delta_{1j} \in_{2kh} \hat{u}_h \hat{u}_l \hat{u}_m \cdots = \left( \frac{\partial^2 B_{13l\cdots}}{\partial x_1 \partial x_2}\hat{u}_1 - \frac{\partial^2 B_{11l\cdots}}{\partial x_1 \partial x_2}\hat{u}_3 \right)\hat{u}_l \hat{u}_m \cdots \quad \text{(C.2-18)}$$

Next let $g=2$, $j=2$, $i=1$. Again summing over $k$ we obtain

$$\frac{\partial^2 B_{jkl\cdots}}{\partial x_2 \partial x_1}\delta_{2j} \in_{1kh} \hat{u}_h \hat{u}_l \hat{u}_m \cdots = \left( \frac{\partial^2 B_{22l\cdots}}{\partial x_2 \partial x_1}\hat{u}_3 - \frac{\partial^2 B_{23l\cdots}}{\partial x_2 \partial x_1}\hat{u}_2 \right)\hat{u}_l \hat{u}_m \cdots \quad \text{(C.2-19)}$$

Employing the mixed-derivative theorem [Eq. (1.3-11)], summing Eqs. (C.2-18) and (C.2-19), and collecting the $\hat{u}_3$ term gives

$$\frac{\partial^2}{\partial x_1 \partial x_2}\left(B_{22l\cdots} - B_{11l\cdots}\right)\hat{u}_3 \hat{u}_l \hat{u}_m \cdots \quad \text{(C.2-20)}$$

It is clear that unless $\frac{\partial^2}{\partial x_1 \partial x_2}\left(B_{22l\cdots} - B_{11l\cdots}\right)$ is zero, the $\hat{u}_3$ term is not zero. Therefore

$$\mathbf{CD}\,\overset{\equiv}{\underset{\cdots}{B}} = \nabla \times \left( \nabla \cdot \overset{\equiv}{\underset{\cdots}{B}} \right) \neq 0 \quad \text{(C.2-21)}$$

in general. This is in contrast to ($\mathbf{DC}\,\overset{\equiv}{\underset{\cdots}{B}} = 0$), which was shown in Section C.2.2.

### C.2.5 Demonstration that the gradient of the curl of any tensor is in general nonzero

Since the first operation is the curl of $\overset{\equiv}{\underset{\cdots}{B}}$, which we have already expressed in Eq. (C.2-8), we may express the gradient of the curl of $\overset{\equiv}{\underset{\cdots}{B}}$ as

$$\nabla\left( \nabla \times \overset{\equiv}{\underset{\cdots}{B}} \right) = \frac{\partial^2 B_{jkl\cdots}}{\partial x_g \partial x_i} \in_{gjh} \hat{u}_i \hat{u}_h \hat{u}_k \hat{u}_m \cdots \quad \text{(C.2-22)}$$

Again, since our objective is to show that the *grad curl* is in general not zero, we need only show that one term is not zero. To do this we arbitrarily pick a combination of indices. If we examine the $\partial x_1 \partial x_2$ and sum over *j*, then we let *g=1*, *i=2*, then

$$\begin{aligned} &\frac{\partial^2 B_{jkl\cdots}}{\partial x_1 \partial x_2} \in_{1jh} \hat{u}_2 \hat{u}_h \hat{u}_k \hat{u}_l \cdots \\ &= \frac{\partial^2 B_{2kl\cdots}}{\partial x_1 \partial x_2} \hat{u}_2 \hat{u}_3 \hat{u}_k \hat{u}_l \cdots - \frac{\partial^2 B_{3kl\cdots}}{\partial x_1 \partial x_2} \hat{u}_2 \hat{u}_2 \hat{u}_k \hat{u}_l \cdots \end{aligned} \tag{C.2-23}$$

Next let *g=2*, *i=1*, then

$$\begin{aligned} &\frac{\partial^2 B_{jkl\cdots}}{\partial x_2 \partial x_1} \in_{2jh} \hat{u}_1 \hat{u}_h \hat{u}_k \hat{u}_l \cdots \\ &= \frac{\partial^2 B_{3kl\cdots}}{\partial x_2 \partial x_1} \hat{u}_1 \hat{u}_1 \hat{u}_k \hat{u}_l \cdots - \frac{\partial^2 B_{1kl\cdots}}{\partial x_2 \partial x_1} \hat{u}_1 \hat{u}_3 \hat{u}_k \hat{u}_l \cdots \end{aligned} \tag{C.2-24}$$

Again it is clear that even after application of the mixed-derivative theorem, cancellation of terms does not take place. Therefore

$$\mathbf{GC}\,\overset{\vdots}{\overline{\overline{\overline{B}}}} = \nabla\left(\nabla \times \overset{\vdots}{\overline{\overline{\overline{B}}}}\right) \neq 0 \tag{C.2-25}$$

in general. This is in contrast to ($\mathbf{CG}\,\overset{\vdots}{\overline{\overline{\overline{B}}}} = 0$), which was shown in Section C.2.3.

### C.2.6 Demonstration of the Lagrange identity applied to tensors.

In Sections 4.7.3 and 4.7.4 the scalar and vector Laplacian operators were developed in GOCCs and cylindrical coordinates. Here we take the opportunity to examine this second-order differential operator in association with general-rank tensors, especially as it relates to the Lagrange identity. This subsection, as in all of Section C.2, is done in Cartesian coordinates for simplicity and with the assurance that there is no loss in generality in using Cartesian coordinates when proving vector and tensor identities. However, there are some issues with Cartesian coordinate differential operator expansions that were addressed in Chapters 1 and 4 and are addressed again in Sections C.3 and C.4.

First we define the Laplacian operator to be the divergence of the gradient of any tensor quantity, which, of course, may include scalars, vectors, or dyadics. This operator may be defined as

$$\nabla^2(\ ) = \nabla \cdot \nabla(\ ) \tag{C.2-26}$$

Since the gradient of a tensor has a resultant rank that is one greater than the rank of the tensor upon which it operates, and the divergence has a resultant one less than its operand, the Laplacian does not change the rank. Thus the Laplacian of a dyadic is a dyadic and the Laplacian of a general-rank tensor is another tensor of the same rank.

After developing the vector Laplacian in Section 4.7.4, which first required the development of the gradient of a vector and then the development of the divergence of the resulting dyadic, the Lagrange vector identity

$$\nabla^2 \overline{A} = \nabla(\nabla \cdot \overline{A}) - \nabla \times \nabla \times \overline{A} \tag{4.7-15}$$

was presented without proof. It was pointed out that this identity is usually presented to undergraduate students as a means of avoiding the dyadic gradient of a vector as well as the divergence of the resulting dyadic. By the use of the right-hand side of Eq. (4.7-15) it is clearly seen that all operations are performed with vector and scalar operands. These are the curl of the curl (**CCv** $= \nabla \times \nabla \times \overline{A}$) and the gradient of the divergence (**GDv** $= \nabla\nabla \cdot \overline{A}$). **CCv** was covered in Section 4.7.5 and **GDv** was developed in Section 4.7.6.

By applying Eqs. (C.2-7) and (C.2-5) to Eq. (C.2-26) we may write the tensor Laplacian as

$$\begin{aligned}
\nabla \cdot \left( \nabla \overset{\vdots}{\overline{\overline{\overline{B}}}} \right) &= \left( \hat{u}_g \frac{\partial}{\partial x_g} \right) \cdot \frac{\partial B_{jkl\cdots}}{\partial x_i} \hat{u}_i \hat{u}_j \hat{u}_k \hat{u}_l \cdots \\
&= \frac{\partial^2 B_{jkl\cdots}}{\partial x_i \partial x_g} \left( \hat{u}_g \cdot \hat{u}_i \right) \hat{u}_j \hat{u}_k \hat{u}_l \cdots \\
&= \frac{\partial^2 B_{jkl\cdots}}{\partial x_i \partial x_g} \delta_{gi} \hat{u}_j \hat{u}_k \hat{u}_l \cdots \\
&= \frac{\partial^2 B_{jkl\cdots}}{\partial x_i^2} \hat{u}_j \hat{u}_k \hat{u}_l \cdots = \left( \nabla \cdot \nabla \right) \overset{\vdots}{\overline{\overline{\overline{B}}}}
\end{aligned} \tag{C.2-27}$$

Thus the expanded form of the left-hand side of Eq. (4.7-15) is

$$\nabla^2 \overset{\dots}{\overline{\overline{\overline{B}}}} = \frac{\partial^2 B_{jkl\cdots}}{\partial x_i^2} \hat{u}_j \hat{u}_k \hat{u}_l \cdots \tag{C.2-28}$$

Next, let us work the right-hand side of Eq. (4.7-15). The first term is the gradient of the divergence, which is

$$\begin{aligned}\nabla(\nabla \cdot \overset{\dots}{\overline{\overline{\overline{B}}}}) &= \left(\hat{u}_g \frac{\partial}{\partial x_g}\right) \frac{\partial B_{jkl\cdots}}{\partial x_j} \hat{u}_k \hat{u}_l \cdots \\ &= \frac{\partial^2 B_{jkl\cdots}}{\partial x_g \partial x_j} \hat{u}_g \hat{u}_k \hat{u}_l \cdots\end{aligned} \tag{C.2-29}$$

and the second term is the negative of the curl of the curl, expanded as

$$\begin{aligned}\nabla \times \left(\nabla \times \overset{\dots}{\overline{\overline{\overline{B}}}}\right) &= \left(\hat{u}_g \frac{\partial}{\partial x_g}\right) \times \frac{\partial B_{jkl\cdots}}{\partial x_i} \in_{ijh} \hat{u}_h \hat{u}_k \hat{u}_l \cdots \\ &= \frac{\partial^2 B_{jkl\cdots}}{\partial x_g \partial x_i} \in_{ijh} \left(\hat{u}_g \times \hat{u}_h\right) \hat{u}_k \hat{u}_l \cdots = \frac{\partial^2 B_{jkl\cdots}}{\partial x_g \partial x_i} \in_{ijh} \in_{ghf} \hat{u}_f \hat{u}_k \hat{u}_l \cdots\end{aligned} \tag{C.2-30}$$

This becomes

$$\begin{aligned}\nabla \times \left(\nabla \times \overset{\dots}{\overline{\overline{\overline{B}}}}\right) &= \left(\hat{u}_g \frac{\partial}{\partial x_g}\right) \times \frac{\partial B_{jkl\cdots}}{\partial x_i} \left(\hat{u}_i \times \hat{u}_h\right) \hat{u}_k \hat{u}_l \cdots \\ &= \frac{\partial^2 B_{jkl\cdots}}{\partial x_g \partial x_i} \left[\hat{u}_g \times \left(\hat{u}_i \times \hat{u}_h\right)\right] \hat{u}_k \hat{u}_l \cdots \\ &= \frac{\partial^2 B_{jkl\cdots}}{\partial x_g \partial x_i} \left[\hat{u}_i \left(\hat{u}_g \cdot \hat{u}_j\right) - \hat{u}_j \left(\hat{u}_g \cdot \hat{u}_i\right)\right] \hat{u}_k \hat{u}_l \cdots \\ &= \frac{\partial^2 B_{jkl\cdots}}{\partial x_g \partial x_i} \left[\delta_{gj} \hat{u}_i - \delta_{gi} \hat{u}_j\right] \hat{u}_k \hat{u}_l \cdots \\ &= \underbrace{\frac{\partial^2 B_{jkl\cdots}}{\partial x_j \partial x_i} \hat{u}_i \hat{u}_k \hat{u}_l \cdots}_{\nabla(\nabla \cdot \overset{\dots}{\overline{\overline{\overline{B}}}})} - \underbrace{\frac{\partial^2 B_{jkl\cdots}}{\partial x_i^2} \hat{u}_j \hat{u}_k \hat{u}_l \cdots}_{\nabla^2 \overset{\dots}{\overline{\overline{\overline{B}}}})}\end{aligned} \tag{C.2-31}$$

Note that the step from the second to the third line of Eq. (C.2-31) made use of the following vector identity

$$\hat{u}_g \times (\hat{u}_i \times \hat{u}_h) = \hat{u}_i (\hat{u}_g \cdot \hat{u}_j) - \hat{u}_j (\hat{u}_g \cdot \hat{u}_i) \tag{C.2-32}$$

The final two terms of the curl curl are seen from Eqs. (C.2-29) and (C.2-28) to be the grad div minus the Laplacian. Rearranging terms, we have

$$\nabla^2 \overset{\cdots}{\overline{\overline{\overline{B}}}} = \nabla(\nabla \cdot \overset{\cdots}{\overline{\overline{\overline{B}}}}) - \nabla \times \nabla \times \overset{\cdots}{\overline{\overline{\overline{B}}}} \tag{C.2-33}$$

and therefore we have shown that the Lagrangian identity may be applied to a general rank tensor.

## C.3 Generalization of the Divergence Operator of Eq. (4.7-7)

The divergence operator of Eq. (4.7-7) was developed inductively and is repeated here for convenience.

$$\nabla \cdot \quad = \frac{1}{h_1 h_2 h_3} \sum_{i=1}^{3} \frac{\partial}{\partial q_i} \left( h_{i+1} h_{i+2} \hat{u}_i \cdot \quad \right) \tag{4.7-7}$$

It was applied to a vector [Eq. (4.4-22)] and a dyadic [Eq. (4.7-9)] and cited for application to any tensor. This section gives a deductive proof of its general application to a tensor. Because of the necessity of accounting for coordinate derivatives of unit vectors, this development is made in generalized orthogonal curvilinear coordinates (GOCCs).

Throughout the main chapters of this guide the importance of dealing with derivatives of unit vectors was emphasized. It was pointed out that although unit vectors do not change in magnitude as coordinate positions change, they can change direction. This fact is important for all coordinates that are curvilinear. Another way to state this is in terms of the Lamé coefficients $h_i$ (also named scale factors and often named metric coefficients). That is, for any coordinate that has a nonunity Lamé coefficient, derivatives of unit vectors with respect to that coordinate will not in general be zero. Eqs. (1.3-19) and (1.3-20) provide the explicit prescription for determining such derivatives. Again these are repeated from page 1-23 for convenience.

$$\frac{\partial \hat{u}_i}{\partial q_i} = -\frac{\hat{u}_j}{h_j}\frac{\partial h_i}{\partial q_j} - \frac{\hat{u}_k}{h_k}\frac{\partial h_i}{\partial q_k} \tag{1.3-19}$$

and

$$\frac{\partial \hat{u}_i}{\partial q_j} = \frac{\hat{u}_j}{h_i}\frac{\partial h_j}{\partial q_i} \tag{1.3-20}$$

where $i = 1,2,3;\ j = 2,3,1$ and $k = 3,1,2$ in that order.

Equations (1.3-19) and (1.3-20) may be combined with the use of the Kronecker delta as

$$\frac{\partial \hat{u}_\alpha}{\partial q_\beta} = \frac{\left(1-\delta_{\alpha\beta}\right)\hat{u}_\beta}{h_\alpha}\frac{\partial h_\beta}{\partial q_\alpha} - \sum_{\gamma=1}^{3}\delta_{\alpha\beta}\left(1-\delta_{\alpha\gamma}\right)\frac{\hat{u}_\gamma}{h_\gamma}\frac{\partial h_\alpha}{\partial q_\gamma} \tag{C.3-1}$$

where $\alpha, \beta, \gamma = 1,2,3$ replace *i, j, k* in (1.3-19) and (1.3-20) above.

The nabla operator in the orthogonal curvilinear coordinates can be written as

$$\nabla = \sum_{i=1}^{3}\frac{\hat{u}_i}{h_i}\frac{\partial}{\partial q_i} \tag{C.3-2}$$

Notice that Eq. (C.3-2) reduces to Cartesian form [Eq. (C.2-1)] when the Lamé coefficients $h_i = 1$. Let us apply this operator to tensor fields.

Our objective in this section is to develop the divergence of a general-rank tensor which is expressed in explicit standard notation as

$$\overset{\vdots}{\overline{\overline{\overline{A}}}} = A_{jkl\cdots}\hat{u}_j\hat{u}_k\hat{u}_l\cdots \tag{C.3-3}$$

where $A_{jkl\cdots}$ are the scalar components of the tensor. The operator in Eq. (C.3-2) may be applied to the general-rank tensor in Eq. (C.3-3) by a *generalized product operator* " $*$ " as a generalized vector differential operator. Thus we have a generalized vector differential operator acting on a general rank tensor in GOCCs as

$$\nabla * \overset{\equiv}{\overline{A}} = \left( \sum_{i=1}^{3} \frac{\hat{u}_i}{h_i} \frac{\partial}{\partial q_i} \right) * \left[ \sum_{j=1}^{3} \sum_{k=1}^{3} \sum_{l=1}^{3} \cdots \left( A_{jkl\cdots} \hat{u}_j \hat{u}_k \hat{u}_l \cdots \right) \right]$$

$$= \sum_{i=1}^{3} \frac{1}{h_i} \sum_{j=1}^{3} \sum_{k=1}^{3} \sum_{l=1}^{3} \cdots \left\{ \frac{\partial A_{jkl\cdots}}{\partial q_i} \left( \hat{u}_i * \hat{u}_j \right) \hat{u}_k \hat{u}_l \ldots \right\} +$$

$$+ \sum_{i=1}^{3} \frac{1}{h_i} \sum_{j=1}^{3} \sum_{k=1}^{3} \sum_{l=1}^{3} \cdots \left\{ A_{jkl\cdots} \left[ \hat{u}_i * \underbrace{\left( \left[ \frac{\left(1-\delta_{ji}\right) \hat{u}_i}{h_j} \frac{\partial h_i}{\partial q_j} - \sum_{g=1}^{3} \delta_{ji} \left(1-\delta_{jg}\right) \frac{\hat{u}_g}{h_g} \frac{\partial h_j}{\partial q_g} \right] \right)}_{\frac{\partial}{\partial q_i}\left(\hat{u}_j\right)} \right] \hat{u}_k \hat{u}_l \ldots \right\}$$

$$+ \sum_{i=1}^{3} \frac{1}{h_i} \sum_{j=1}^{3} \sum_{k=1}^{3} \sum_{l=1}^{3} \cdots \left[ A_{jkl\cdots} \left( \hat{u}_i * \hat{u}_j \right) \left( \frac{\partial \hat{u}_k}{\partial q_i} \hat{u}_l \cdots \quad + \hat{u}_k \frac{\partial \hat{u}_l}{\partial q_i} \hat{u}_m \cdots \quad + \cdots \right) \right]$$

(C.3-4)

where the sign * denotes dot ( $\cdot$ ), cross ( $\times$ ) or tensor ( $\otimes$ ) multiplication.

*Although the nabla vector operator of Eq. (C.3-2) has similarities with a vector when performing the product forms above, there are substantive differences. It is of paramount importance to stress that the derivative must be taken not only on the tensor component* $A_{jkl\cdots}$ *and the first unit vector* $\hat{u}_j$ *but also on all of the remaining unit vectors as well. If the nabla operator were treated as if it were just a vector, the expression written above would contain the two curly bracketed terms {} only.*

This is more precisely demonstrated in Section C.4. Notice that the bracketed factor in the third term of Eq. (C.3-4) is replaced by Eq. (C.3-1).

Whereas Eq. (C.3-4) can be used for curl and gradient applications, we make use of it here only for the divergence application. Therefore, the divergence of a general-rank tensor in GOCCs is

$$\nabla\cdot\overset{\equiv}{\overset{\cdots}{A}}=\sum_{i=1}^{3}\frac{1}{h_i}\sum_{j=1}^{3}\sum_{k=1}^{3}\sum_{l=1}^{3}\cdots\left\{\frac{\partial A_{jkl\cdots}}{\partial q_j}\delta_{ij}\hat{u}_k\hat{u}_l\ldots\right\}$$
$$+\sum_{i=1}^{3}\frac{1}{h_i}\sum_{j=1}^{3}\sum_{k=1}^{3}\sum_{l=1}^{3}\cdots\left\{A_{jkl\cdots}\left[\begin{array}{l}\dfrac{\left(1-\delta_{ji}\right)}{h_j}\dfrac{\partial h_i}{\partial q_j}-\\ \displaystyle\sum_{g=1}^{3}\underbrace{\delta_{ji}\left(1-\delta_{jg}\right)\delta_{ig}}_{=0}\frac{1}{h_g}\frac{\partial h_j}{\partial q_g}\end{array}\right]\hat{u}_k\hat{u}_l\ldots\right\} \tag{C.3-5}$$
$$+\sum_{i=1}^{3}\frac{1}{h_i}\sum_{j=1}^{3}\sum_{k=1}^{3}\sum_{l=1}^{3}\left[A_{jkl\cdots}\delta_{ij}\left(\frac{\partial\hat{u}_k}{\partial q_i}\hat{u}_l\cdots+\hat{u}_k\frac{\partial\hat{u}_l}{\partial q_i}\hat{u}_m\cdots+\cdots\right)\right]$$

A well-known vector identity

$$\delta_{ji}\left(1-\delta_{jg}\right)\delta_{ig}=0 \tag{C.3-6}$$

may readily be checked by testing all possible combinations of indices. Thus, the second term in the second curly brackets is zero as shown. In addition the Kronecker delta in the first and third terms eliminates the $i$ summation signs. Therefore Eq. (C.3-5) becomes

$$\nabla\cdot\overset{\equiv}{\overset{\cdots}{A}}=\sum_{j=1}^{3}\frac{1}{h_j}\sum_{k=1}^{3}\sum_{l=1}^{3}\cdots\left(\frac{\partial A_{jkl\cdots}}{\partial q_j}\hat{u}_k\hat{u}_l\ldots\right)$$
$$+\sum_{j=1}^{3}\sum_{k=1}^{3}\sum_{l=1}^{3}\cdots\left[A_{jkl\cdots}\underbrace{\sum_{i=1}^{3}\frac{\left(1-\delta_{ji}\right)}{h_ih_j}\frac{\partial h_i}{\partial q_j}}_{\left(\frac{1}{h_jh_{j+1}}\frac{\partial h_{j+1}}{\partial q_j}+\frac{1}{h_jh_{j+2}}\frac{\partial h_{j+2}}{\partial q_j}\right)}\hat{u}_k\hat{u}_l\ldots\right] \tag{C.3-7}$$
$$+\sum_{j=1}^{3}\frac{1}{h_i}\sum_{k=1}^{3}\sum_{l=1}^{3}\left[A_{jkl\cdots}\left(\frac{\partial\hat{u}_k}{\partial q_i}\hat{u}_l\cdots+\hat{u}_k\frac{\partial\hat{u}_l}{\partial q_i}\hat{u}_m\cdots+\cdots\right)\right]$$

Since Eq. (4.7-7) uses the index rolling rule, namely when $j=1,2,3$, $j+1=2,3,1$, and $j+2=3,1,2$, respectively, we can express the derivative factor in the second term of Eq. (C.3-7) as shown. Further noting from the rule of derivatives of multiple variables that

$$\left(\frac{1}{h_j h_{j+1}}\frac{\partial h_{j+1}}{\partial q_j}+\frac{1}{h_j h_{j+2}}\frac{\partial h_{j+2}}{\partial q_j}\right)=\frac{1}{h_j h_{j+1} h_{j+2}}\frac{\partial\left(h_{j+1}h_{j+2}\right)}{\partial q_j}$$
$$=\frac{1}{h_1 h_2 h_3}\frac{\partial\left(h_{j+1}h_{j+2}\right)}{\partial q_j} \tag{C.3-8}$$

and thus Eq. (C.3-7) becomes

$$\nabla\cdot\overset{\equiv}{A}=\sum_{j=1}^{3}\sum_{k=1}^{3}\sum_{l=1}^{3}\cdots\left[\frac{1}{h_j}\frac{\partial A_{jkl\cdots}}{\partial q_j}\hat{u}_k\hat{u}_l\cdots\ +A_{jkl\cdots}\frac{1}{h_1h_2h_3}\frac{\partial\left(h_{j+1}h_{j+2}\right)}{\partial q_j}\hat{u}_k\hat{u}_l\cdots\right]$$
$$+\frac{1}{h_j}\left[A_{jkl\cdots}\frac{\partial\hat{u}_k}{\partial q_j}\hat{u}_l\cdots\ +A_{jkl\cdots}\hat{u}_k\frac{\partial\hat{u}_l}{\partial q_j}\hat{u}_m\cdots\ +\cdots\right] \tag{C.3-9}$$

Again from the rule of derivatives of multiple variables all three terms of Eq. (C.3-9) may be combined simplifying the divergence of $\overset{\equiv}{A}$ as

$$\nabla\cdot\overset{\equiv}{A}=\sum_{j=1}^{3}\sum_{k=1}^{3}\sum_{l=1}^{3}\cdots\left[\frac{1}{h_1h_2h_3}\frac{\partial\left(A_{jkl\cdots}h_{j+1}h_{j+2}\hat{u}_k\hat{u}_l\cdots\right)}{\partial q_j}\right]$$
$$=\frac{1}{h_1}\frac{1}{h_2}\frac{1}{h_3}\sum_{j=1}^{3}\sum_{k=1}^{3}\sum_{l=1}^{3}\cdots\left[\frac{\partial\left(\hat{u}_j\cdot A_{jkl\cdots}h_{j+1}h_{j+2}\hat{u}_j\hat{u}_k\hat{u}_l\cdots\right)}{\partial q_j}\right]$$
$$=\frac{1}{h_1}\frac{1}{h_2}\frac{1}{h_3}\sum_{j=1}^{3}\sum_{k=1}^{3}\sum_{l=1}^{3}\cdots\left[\frac{\partial\left(\sum_{i=1}^{3}\delta_{ij}\hat{u}_i\cdot A_{jkl\cdots}h_{j+1}h_{j+2}\hat{u}_j\hat{u}_k\hat{u}_l\cdots\right)}{\partial q_j}\right]$$
$$=\frac{1}{h_1}\frac{1}{h_2}\frac{1}{h_3}\sum_{i=1}^{3}\left[\frac{\partial\left(h_{i+1}h_{i+2}\hat{u}_i\cdot\sum_{j=1}^{3}\sum_{k=1}^{3}\sum_{l=1}^{3}\cdots A_{jkl\cdots}\hat{u}_j\hat{u}_k\hat{u}_l\cdots\right)}{\partial q_i}\right]$$
$$=\frac{1}{h_1}\frac{1}{h_2}\frac{1}{h_3}\sum_{i=1}^{3}\frac{\partial}{\partial q_i}\left(h_{i+1}h_{i+2}\hat{u}_i\cdot\overset{\equiv}{A}\right) \tag{C.3-10}$$

At the third equality above, a dummy index $i$ was introduced. In prior sections of this appendix the index $g$ was used for this purpose, however, to demonstrate the generality of Eq. (4.7-7), which contains the $i$, we use this as our dummy index. This should not be confused with the $i$ index in Eqs. (C.3-4) and (C.3-5).

Notice also that the last equality is precisely the form of Eq. (4.7-7) and thus we have our QED (quod erat demonstrandum). Namely we have shown that the divergence operator of Eq. (4.7-7) is applicable to tensors of general rank. The last equality is another equivalent form of Eq. (4.7-7).

## C.4 The Dual Nature of the Nabla Operator

The generalized form of the vector Laplacian [Eq. (4.7-11)] was obtained by first taking the gradient of the vector and then taking the divergence of the resulting dyadic. The Lagrange vector identity, Eq. (4.7-15), provided a means of determining the vector Laplacian without resorting to dyadics; however it is customarily cited without proof. This identity was proven for generalized rank tensors in Section C.2.6. Although performed in Cartesian coordinates, its validity is independent of the coordinate system.

The process of demonstrating such relationships in Cartesian coordinates has led many to believe that the nabla operators ($\nabla\cdot, \nabla\times$, and $\nabla$) are analogous to the vector dot-, cross- and direct-product[3] operations ($\bar{A}\cdot$, $\bar{A}\times$, and $\bar{A}$). Such a supposition is very dangerous and will lead to erroneous results in all but Cartesian coordinates.

Therefore, we find that it is necessary to take into account the derivative nature of the nabla operator and not just its dot-, cross-, or direct-product nature. The necessity of using both characteristics of the nabla operator is brought out quite effectively by proving the Lagrange vector identity in generalized orthogonal curvilinear coordinates (GOCCs). That is, the demonstration that the vector Laplacian as determined from the l.h.s. of the Lagrange identity is equated to the vector Laplacian from the r.h.s. in GOCCs requires the use of both natures of the nabla operator. Confirmation of this in GOCCs proved to be quite protracted. So, in the interest of brevity we omit this proof (of the Lagrange vector identity in GOCCs) and simply show that the dot, cross, and direct product of a vector with a general-rank tensor yields different resultant forms from the counterpart nabla operations on the same tensor.

Again, as introduced in Section C.1, the generalized-product operator—the asterisk "$*$" operator—is used to denote either a dot-, cross-, or tensor-product operation. We show that extra terms come into play with the $\nabla * \overset{\cdots}{\overline{\overline{\overline{T}}}}$ operation as compared with the $\overline{A} * \overset{\cdots}{\overline{\overline{\overline{T}}}}$ operation. In particular the latter operation takes the form

$$\begin{aligned}\overline{A} * \overset{\cdots}{\overline{\overline{\overline{T}}}} &= \sum_{i=1}^{3} A_i \hat{u}_i * \sum_{j=1}^{3}\sum_{k=1}^{3}\sum_{l=1}^{3} \cdots T_{jkl\cdots} \hat{u}_j \hat{u}_k \hat{u}_l \cdots \\ &= \sum_{i=1}^{3}\sum_{j=1}^{3}\sum_{k=1}^{3} \cdots A_i T_{jkl\cdots} \left(\hat{u}_i * \hat{u}_j\right) \hat{u}_k \hat{u}_l \cdots\end{aligned} \qquad \text{(C.4-1)}$$

In contrast to Eq. (C.4-1) the generalized-nabla-product operating on the same tensor takes the form

$$\begin{aligned}\nabla * \overset{\cdots}{\overline{\overline{\overline{T}}}} &= \sum_{i=1}^{3} \left( \frac{\hat{u}_i}{h_i} \frac{\partial}{\partial q_i} \right) * \left( \sum_{j=1}^{3}\sum_{k=1}^{3}\sum_{l=1}^{3} \cdots T_{jkl\cdots} \hat{u}_j \hat{u}_k \hat{u}_l \cdots \right) \\ &= \sum_{i=1}^{3}\sum_{j=1}^{3}\sum_{k=1}^{3} \cdots \frac{\partial T_{jkl\cdots}}{h_i \partial q_i} \left(\hat{u}_i * \hat{u}_j\right) \hat{u}_k \hat{u}_l \cdots \\ &+ \sum_{i=1}^{3}\sum_{j=1}^{3}\sum_{k=1}^{3} \cdots T_{jkl\cdots} \left( \hat{u}_i * \frac{\partial \hat{u}_j}{h_i \partial q_i} \right) \hat{u}_k \hat{u}_l \cdots \\ &+ \sum_{i=1}^{3}\sum_{j=1}^{3}\sum_{k=1}^{3} \cdots T_{jkl\cdots} \left(\hat{u}_i * \hat{u}_j\right) \left( \frac{\partial \hat{u}_k}{h_i \partial q_i} \right) \hat{u}_l \cdots \\ &+ \cdots\end{aligned} \qquad \text{(C.4-2)}$$

Notice that the first term of Eq. (C.4-2) is analogous to Eq. (C.4-1) but that the second, third and subsequent terms are entirely missed in Eq. (C.4-1). Therefore, the popular statement that nabla operator is analogous to the vector operator is not valid in cases where these latter terms come into play.

To illustrate where it is valid and where it is not, let us take the case of a rank-one operand, namely, let $\overset{\cdots}{\overline{\overline{\overline{T}}}}$ be the vector $\overline{B}$. Then Eq. (C.4-1) becomes

$$\bar{A} * \bar{B} = \sum_{i=1}^{3} A_i \hat{u}_i * \sum_{j=1}^{3} B_j \hat{u}_j = \sum_{i=1}^{3} \sum_{j=1}^{3} A_i B_j \left( \hat{u}_i * \hat{u}_j \right) \tag{C.4-3}$$

and Eq. (C.4-2) becomes

$$\nabla * \bar{B} = \sum_{i=1}^{3} \left( \frac{\hat{u}_i}{h_i} \frac{\partial}{\partial q_i} \right) * \left( \sum_{j=1}^{3} B_j \hat{u}_j \right) = \sum_{i=1}^{3} \sum_{j=1}^{3} \frac{\partial B_j}{h_i \partial q_i} \left( \hat{u}_i * \hat{u}_j \right) + \sum_{i=1}^{3} \sum_{j=1}^{3} B_j \left( \hat{u}_i * \frac{\partial \hat{u}_j}{h_i \partial q_i} \right) \tag{C.4-4}$$

It is clear that forms (C.4-3) and (C.4-4) are not analogous unless all of the coordinate derivatives of unit vectors $\partial \hat{u}_j / \partial q_i$ are zero. This is the case only in Cartesian coordinates.

For the case of a rank-two operand, we let $\overset{\dots}{\overline{\overline{\overline{T}}}}$ be the dyadic $\overline{\overline{D}}$. Then Eq. (C.4-1) becomes

$$\bar{A} * \overline{\overline{D}} = \sum_{i=1}^{3} A_i \hat{u}_i * \sum_{j=1}^{3} \sum_{k=1}^{3} D_{jk} \hat{u}_j \hat{u}_k = \sum_{i=1}^{3} \sum_{j=1}^{3} \sum_{k=1}^{3} A_i D_{jk} \left( \hat{u}_i * \hat{u}_j \right) \hat{u}_k \tag{C.4-5}$$

and Eq. (C.4-2) becomes

$$\begin{aligned}
\nabla * \bar{\bar{D}} &= \sum_{i=1}^{3}\left(\frac{\hat{u}_i}{h_i}\frac{\partial}{\partial q_i}\right) * \left(\sum_{j=1}^{3}\sum_{k=1}^{3} D_{jk}\hat{u}_j\hat{u}_k\right) \\
&= \sum_{i=1}^{3}\sum_{j=1}^{3}\sum_{k=1}^{3}\frac{\partial D_{jk}}{h_i \partial q_i}\left(\hat{u}_i * \hat{u}_j\right)\hat{u}_k \\
&+ \sum_{i=1}^{3}\sum_{j=1}^{3}\sum_{k=1}^{3} D_{jk}\left(\hat{u}_i * \frac{\partial \hat{u}_j}{h_i \partial q_i}\right)\hat{u}_k \\
&+ \sum_{i=1}^{3}\sum_{j=1}^{3}\sum_{k=1}^{3} D_{jk}\left(\hat{u}_i * \hat{u}_j \frac{\partial \hat{u}_k}{h_i \partial q_i}\right)
\end{aligned} \qquad \text{(C.4-6)}$$

Again the analogy does not hold unless the second and third terms go to zero. In Cartesian coordinates the unit vector coordinate derivatives in these terms will be zero and, therefore, the analogy holds in this case. In any other coordinate system, it does not. By induction the same conclusion can be made for all higher-rank tensors. Therefore, we have demonstrated the following axioms:

*Nabla operators have two natures: a product nature and a derivative nature.*

*Vector operators have one nature: product nature.*

*Nabla operators are not in general analogous to vector operators, except in Cartesian coordinates.*

*In Cartesian coordinates, nabla operators are analogous to vector operators.*

## Reference

1. David C. Kay, "The Einstein Summation Convention," in *Theory and Problems of Tensor Calculus*, Schaum's Outline Series, McGraw-Hill (1988).

## Appendix D

# Coordinate Expansions of Vector Differential Operators

This appendix provides Cartesian and cylindrical coordinate expansions of first- and second-order vector differential operators acting on scalar, vector, and dyadic operands. The divergence and curl of vector and dyadic fields and the gradient of scalar, vector, and dyadic fields are provided with applications cited for the curl of the strain dyadic and the gradient of the stress dyadic. In addition the scalar and vector Laplacian expansions are listed.

## D.1 Cartesian Coordinate Expansions

The conversion from generalized orthogonal curvilinear coordinates (GOCCs) to Cartesian coordinates is

$$q_1 = x\,,\; q_2 = y\,,\; q_3 = z \qquad \text{(D.1-1a,b,c)}$$

and

$$h_1 = 1\,,\; h_2 = 1\,,\; h_3 = 1 \qquad \text{(D.1-2a,b,c)}$$

### D.1.1 Cartesian coordinate expansions of first-order vector differential operators

First-order vector differential operators are expanded in Cartesian coordinates in the order of increasing resultant rank. Thus divergence is done first because the resultant rank is one less than the rank of the operand. Next, the curl is shown because the resultant rank is unchanged. Lastly, the gradient is displayed because the resultant rank is increased by one.

#### D.1.1(a) The divergence of vector and dyadic fields

The divergence of a vector field [from Eq. (4.4-22)] is the scalar field

$$\nabla \cdot \bar{A}\Big|_{Cartesian} = \frac{\partial A_x}{\partial x} + \frac{\partial A_y}{\partial y} + \frac{\partial A_z}{\partial z} \tag{D.1-3}$$

The divergence of a dyadic field [ref: Eq. (B.1-5)] is the vector field

$$\nabla \cdot \bar{\bar{G}}\Big|_{Cartesian} = \hat{u}_x \left[ \frac{\partial G_{xx}}{\partial x} + \frac{\partial G_{yx}}{\partial y} + \frac{\partial G_{zx}}{\partial z} \right] + \hat{u}_y \left[ \frac{\partial G_{xy}}{\partial x} + \frac{\partial G_{yy}}{\partial y} + \frac{\partial G_{zy}}{\partial z} \right] + \hat{u}_z \left[ \frac{\partial G_{xz}}{\partial x} + \frac{\partial G_{yz}}{\partial y} + \frac{\partial G_{zz}}{\partial z} \right] \tag{D.1-4}$$

## D.1.1(b) The curl of vector and dyadic fields

The curl of a vector field $\bar{A}$ [from Eq. (4.5-12)] is the vector field

$$\nabla \times \bar{A}\Big|_{Cartesian} = \hat{u}_x \left( \frac{\partial A_z}{\partial y} - \frac{\partial A_y}{\partial z} \right) + \hat{u}_y \left( \frac{\partial A_x}{\partial z} - \frac{\partial A_z}{\partial x} \right) + \hat{u}_z \left( \frac{\partial A_y}{\partial x} - \frac{\partial A_x}{\partial y} \right) \tag{D.1-5}$$

*Application: The curl of the strain dyadic*

The curl of a dyadic field is useful in advanced studies of mechanics of solids. For example, the displacement vector $\bar{d}$ is determined by integrating the tangential component of the curl of the strain over the path, namely $\bar{d} = \int \left( \nabla \times \bar{\bar{\delta}} \right) \cdot d\bar{\ell}$. Since the resultant of the curl of a dyadic is another dyadic, this line integral takes the form of Eq. (5.1-4), and is thus an application of that line-integral form.

Here we display the curl of the strain dyadic $\bar{\bar{\delta}}$ in Cartesian coordinates as an example of the curl of a dyadic field in general.

$$\begin{aligned} \nabla \times \bar{\bar{\delta}}\Big|_{Cartesian} &= \left( \hat{u}_y \frac{\partial}{\partial z} - \hat{u}_z \frac{\partial}{\partial y} \right) \left( \delta_{xx}\hat{u}_x + \delta_{xy}\hat{u}_y + \delta_{xz}\hat{u}_z \right) \\ &+ \left( \hat{u}_z \frac{\partial}{\partial x} - \hat{u}_x \frac{\partial}{\partial z} \right) \left( \delta_{yx}\hat{u}_x + \delta_{yy}\hat{u}_y + \delta_{yz}\hat{u}_z \right) \\ &+ \left( \hat{u}_x \frac{\partial}{\partial y} - \hat{u}_y \frac{\partial}{\partial x} \right) \left( \delta_{zx}\hat{u}_x + \delta_{zy}\hat{u}_y + \delta_{zz}\hat{u}_z \right) \end{aligned} \tag{D.1-6}$$

which, when expanded [using Eq. (3.3-4)], becomes the dyadic with its nine components as

$$\begin{aligned}\nabla\times\overline{\overline{\delta}}\Big|_{Cart} &= \hat{\hat{u}}_{xx}\left[\frac{\partial\delta_{zx}}{\partial y}-\frac{\partial\delta_{yx}}{\partial z}\right]+\hat{\hat{u}}_{xy}\left[\frac{\partial\delta_{zy}}{\partial y}-\frac{\partial\delta_{yy}}{\partial z}\right]+\hat{\hat{u}}_{xz}\left[\frac{\partial\delta_{zz}}{\partial y}-\frac{\partial\delta_{yz}}{\partial z}\right]\\ &+\hat{\hat{u}}_{yx}\left[\frac{\partial\delta_{xx}}{\partial z}-\frac{\partial\delta_{zx}}{\partial x}\right]+\hat{\hat{u}}_{yy}\left[\frac{\partial\delta_{xy}}{\partial z}-\frac{\partial\delta_{zy}}{\partial x}\right]+\hat{\hat{u}}_{yz}\left[\frac{\partial\delta_{xz}}{\partial z}-\frac{\partial\delta_{zz}}{\partial x}\right]\\ &+\hat{\hat{u}}_{zx}\left[\frac{\partial\delta_{yx}}{\partial x}-\frac{\partial\delta_{xx}}{\partial y}\right]+\hat{\hat{u}}_{zy}\left[\frac{\partial\delta_{yy}}{\partial x}-\frac{\partial\delta_{xy}}{\partial y}\right]+\hat{\hat{u}}_{zz}\left[\frac{\partial\delta_{yz}}{\partial x}-\frac{\partial\delta_{xz}}{\partial y}\right]\end{aligned} \quad \text{(D.1-7)}$$

### D.1.1(c) The gradient of scalar, vector, and dyadic fields

The gradient of a scalar field [from Eq. (4.3-18)] is the vector field

$$\nabla V\Big|_{Cartesian} = \hat{u}_x\frac{\partial V}{\partial x}+\hat{u}_y\frac{\partial V}{\partial y}+\hat{u}_z\frac{\partial V}{\partial z} \quad \text{(D.1-8)}$$

In Section 4.3 we determined the vector Laplacian by first finding the gradient of a vector field in GOCCs. Here we expand that operation in Cartesian coordinates. Thus the gradient of a vector field from Eq. (4.3-20) is the nine-term dyadic field

$$\begin{aligned}\nabla\overline{A}\Big|_{Cartesian} &= \hat{\hat{u}}_{xx}\frac{\partial A_x}{\partial x}+\hat{\hat{u}}_{xy}\frac{\partial A_y}{\partial x}+\hat{\hat{u}}_{xz}\frac{\partial A_z}{\partial x}\\ &+\hat{\hat{u}}_{yx}\frac{\partial A_x}{\partial y}+\hat{\hat{u}}_{yy}\frac{\partial A_y}{\partial y}+\hat{\hat{u}}_{yz}\frac{\partial A_z}{\partial y}\\ &+\hat{\hat{u}}_{zx}\frac{\partial A_x}{\partial z}+\hat{\hat{u}}_{zy}\frac{\partial A_y}{\partial z}+\hat{\hat{u}}_{zz}\frac{\partial A_z}{\partial z}\end{aligned} \quad \text{(D.1-9)}$$

*Application: The gradient of the stress dyadic*

In advanced mechanics of solids, the theory of moments in general necessitates the determination of the gradient of the stress dyadic $\overline{\overline{s}}$. Therefore, we show the expansion of the gradient of the stress dyadic as an application for the gradient of a dyadic field in general.

The gradient of a dyadic field[1] is the 27-term triadic field

$$
\begin{aligned}
\nabla\overline{\overline{s}}\Big|_{cyl} = {} & \hat{\hat{\hat{u}}}_{xxx}\frac{\partial s_{xx}}{\partial x} + \hat{\hat{\hat{u}}}_{xxy}\frac{\partial s_{xy}}{\partial x} + \hat{\hat{\hat{u}}}_{xxz}\frac{\partial s_{xz}}{\partial x} \\
& + \hat{\hat{\hat{u}}}_{xyx}\frac{\partial s_{yx}}{\partial x} + \hat{\hat{\hat{u}}}_{xyy}\frac{\partial s_{yy}}{\partial x} + \hat{\hat{\hat{u}}}_{xyz}\frac{\partial s_{yz}}{\partial x} \\
& + \hat{\hat{\hat{u}}}_{xzx}\frac{\partial s_{zx}}{\partial x} + \hat{\hat{\hat{u}}}_{xzy}\frac{\partial s_{zy}}{\partial x} + \hat{\hat{\hat{u}}}_{xzz}\frac{\partial s_{zz}}{\partial x} \\
& + \hat{\hat{\hat{u}}}_{yxx}\frac{\partial s_{xx}}{\partial y} + \hat{\hat{\hat{u}}}_{yxy}\frac{\partial s_{xy}}{\partial y} + \hat{\hat{\hat{u}}}_{yxz}\frac{\partial s_{xz}}{\partial y} \\
& + \hat{\hat{\hat{u}}}_{yyx}\frac{\partial s_{yx}}{\partial y} + \hat{\hat{\hat{u}}}_{yyy}\frac{\partial s_{yy}}{\partial y} + \hat{\hat{\hat{u}}}_{yyz}\frac{\partial s_{yz}}{\partial y} \\
& + \hat{\hat{\hat{u}}}_{yzx}\frac{\partial s_{zx}}{\partial y} + \hat{\hat{\hat{u}}}_{yzy}\frac{\partial s_{zy}}{\partial y} + \hat{\hat{\hat{u}}}_{yzz}\frac{\partial s_{zz}}{\partial y} \\
& + \hat{\hat{\hat{u}}}_{zxx}\frac{\partial s_{xx}}{\partial z} + \hat{\hat{\hat{u}}}_{zxy}\frac{\partial s_{xy}}{\partial z} + \hat{\hat{\hat{u}}}_{zxz}\frac{\partial s_{xz}}{\partial z} \\
& + \hat{\hat{\hat{u}}}_{zyx}\frac{\partial s_{yx}}{\partial z} + \hat{\hat{\hat{u}}}_{zyy}\frac{\partial s_{yy}}{\partial z} + \hat{\hat{\hat{u}}}_{zyz}\frac{\partial s_{yz}}{\partial z} \\
& + \hat{\hat{\hat{u}}}_{zzx}\frac{\partial s_{zx}}{\partial z} + \hat{\hat{\hat{u}}}_{zzy}\frac{\partial s_{zy}}{\partial z} + \hat{\hat{\hat{u}}}_{zzz}\frac{\partial s_{zz}}{\partial z}
\end{aligned}
\qquad \text{(D.1-10)}
$$

## D.1.2 Cartesian coordinate expansions of second-order vector differential operators

### D.1.2(a) The scalar and vector Laplacian

The scalar Laplacian from Eq. (4.7-4) is the scalar field

$$\nabla^2 V\Big|_{Cartesian} = \nabla \cdot \nabla V = \frac{\partial^2 V}{\partial x^2} + \frac{\partial^2 V}{\partial y^2} + \frac{\partial^2 V}{\partial z^2} \qquad \text{(D.1-11)}$$

The vector Laplacian from Eq. (4.7-11) is the vector field

$$\nabla^2 \overline{A}\Big|_{Cartesian} = \nabla \cdot \nabla \overline{A} = \hat{u}_x \nabla^2 A_x + \hat{u}_y \nabla^2 A_y + \hat{u}_z \nabla^2 A_z \qquad \text{(D.1-12)}$$

### D.1.2(b) The curl of the curl of a vector field

In Section 4.7.5 the curl of the curl of a vector field was determined in order to provide the reader with a methodology of determining the vector Laplacian without resorting to dyadic operations. That is, with the use of the right-hand side of the Lagrange identity [Eq. (4.7-15)], the curl of the curl of a vector field was needed. Here we provided the curl of the curl of a vector field from Eq. (4.7-23), which is the vector field

$$\nabla\times\nabla\times\overline{A}\Big|_{Cart}=\hat{u}_x\left[\left(\frac{\partial^2 A_y}{\partial y\partial x}-\frac{\partial^2 A_x}{\partial y^2}\right)-\left(\frac{\partial^2 A_x}{\partial z^2}-\frac{\partial^2 A_z}{\partial z\partial x}\right)\right]$$
$$+\hat{u}_y\left[\left(\frac{\partial^2 A_z}{\partial z\partial y}-\frac{\partial^2 A_y}{\partial z^2}\right)-\left(\frac{\partial^2 A_y}{\partial x^2}-\frac{\partial^2 A_x}{\partial x\partial y}\right)\right] \quad (D.1\text{-}13)$$
$$+\hat{u}_z\left[\left(\frac{\partial^2 A_x}{\partial x\partial z}-\frac{\partial^2 A_z}{\partial x^2}\right)-\left(\frac{\partial^2 A_z}{\partial y^2}-\frac{\partial^2 A_y}{\partial y\partial z}\right)\right]$$

### D.1.2(c) The gradient of the divergence

In Section 4.7.6 other terms in the right-hand side of the Lagrange identity, namely the gradient of the divergence, were needed to complete the exercise. Here we provide the Cartesian expansion of the gradient of the divergence of a vector field from Eq. (4.7-24):

$$\nabla\nabla\cdot\overline{A}\Big|_{Cartesian}=\hat{u}_x\left(\frac{\partial^2 A_x}{\partial x^2}+\frac{\partial^2 A_y}{\partial x\partial y}+\frac{\partial^2 A_z}{\partial x\partial z}\right)$$
$$+\hat{u}_y\left(\frac{\partial^2 A_x}{\partial y\partial x}+\frac{\partial^2 A_y}{\partial y^2}+\frac{\partial^2 A_z}{\partial y\partial z}\right)+\hat{u}_z\left(\frac{\partial^2 A_x}{\partial z\partial x}+\frac{\partial^2 A_y}{\partial z\partial y}+\frac{\partial^2 A_z}{\partial z^2}\right) \quad (D.1\text{-}14)$$

which again is a vector field.

The inside front cover provides some of the more common first- and second-order vector differential operator expansions in Cartesian coordinates for the reader's fingertip reference.

## D.2 Cylindrical Coordinate Expansions

The conversion from generalized orthogonal curvilinear coordinates (GOCCs) to cylindrical coordinates is

$$q_1=r,\ q_2=\phi,\ q_3=z \quad (D.2\text{-}1a,b,c)$$

and

$$h_1 = 1\,,\; h_2 = r\,,\; h_3 = 1 \qquad \text{(D.2-2a,b,c)}$$

### D.2.1 Cylindrical coordinate expansions of first-order vector differential operators

First-order vector differential operators are expanded in cylindrical coordinates in the order of increasing resultant rank. As in the Cartesian expansions, the divergence is done first because the resultant rank is one less than the rank of the operand, followed by the curl because the resultant rank is unchanged, and then the gradient because the resultant rank is increased by one.

#### D.2.1(a) The divergence of vector and dyadic fields

The divergence of a vector field [from Eq. (4.4-22)] is the scalar field

$$\nabla\cdot\bar{A}\Big|_{cyl} = \frac{1}{r}\frac{\partial(rA_r)}{\partial r} + \frac{1}{r}\frac{\partial A_\phi}{\partial\phi} + \frac{\partial A_z}{\partial z} \qquad \text{(D.2-3)}$$

The divergence of a dyadic field [ref: Eq. (B.3-11)] is the vector field

$$\nabla\cdot\bar{\bar{G}}\Big|_{cyl} = \hat{u}_r\frac{1}{r}\left[\frac{\partial(rG_{rr})}{\partial r} + \frac{\partial G_{\phi r}}{\partial\phi} - G_{\phi\phi} + r\frac{\partial G_{zr}}{\partial z}\right]$$
$$+\hat{u}_\phi\frac{1}{r}\left[\frac{\partial(rG_{r\phi})}{\partial r} + \frac{\partial G_{\phi\phi}}{\partial\phi} + G_{\phi r} + r\frac{\partial G_{z\phi}}{\partial z}\right] + \hat{u}_z\frac{1}{r}\left[\frac{\partial(rG_{rz})}{\partial r} + \frac{\partial G_{\phi z}}{\partial\phi} + r\frac{\partial G_{zz}}{\partial z}\right]$$
(D.2-4)

#### D.2.1(b) The curl of vector and dyadic fields

The curl of a vector field [ref: Eq. (4.5-13)] is the vector field

$$\nabla\times\bar{A}\Big|_{cyl} = \hat{u}_r\left[\frac{1}{r}\frac{\partial A_z}{\partial\phi} - \frac{\partial A_\phi}{\partial z}\right] + \hat{u}_\phi\left[\frac{\partial A_r}{\partial z} - \frac{\partial A_z}{\partial r}\right] + \hat{u}_z\frac{1}{r}\left[\frac{\partial(rA_\phi)}{\partial r} - \frac{\partial A_r}{\partial\phi}\right] \qquad \text{(D.2-5)}$$

The curl of a dyadic field [ref: Eq. (B.2-5)] is the dyadic field

$$\nabla\times\overline{\overline{G}}\Big|_{cyl}=\left(\hat{u}_\phi\frac{\partial}{\partial z}-\hat{u}_z\frac{1}{r}\frac{\partial}{\partial\phi}\right)\left(G_{rr}\hat{u}_r+G_{r\phi}\hat{u}_\phi+G_{rz}\hat{u}_z\right)$$
$$+\left(\hat{u}_z\frac{1}{r}\frac{\partial}{\partial r}-\hat{u}_r\frac{1}{r}\frac{\partial}{\partial z}\right)\left(rG_{\phi r}\hat{u}_r+rG_{\phi\phi}\hat{u}_\phi+rG_{\phi z}\hat{u}_z\right) \qquad \text{(D.2-6)}$$
$$+\left(\hat{u}_r\frac{1}{r}\frac{\partial}{\partial\phi}-\hat{u}_\phi\frac{\partial}{\partial r}\right)\left(G_{zr}\hat{u}_r+G_{z\phi}\hat{u}_\phi+G_{zz}\hat{u}_z\right)$$

[Notice that by replacing the $r$ and $\phi$ coordinates with $x$ and $y$, and by replacing (D.2-2b) with (D.1-2b), that is, by letting the $r$ and $1/r$ coefficients become unity, Equations (D.2-1) through (D.2-6) become (D.1-1) through (D.1-6).]

When the derivative operators of (D.2-6) are taken, the curl of the dyadic [using Eq. (3.3-4)] becomes the nine-dyadic component expansion

$$\nabla\times\overline{\overline{G}}\Big|_{cyl}=\hat{\hat{u}}_{rr}\left[\frac{1}{r}\frac{\partial G_{zr}}{\partial\phi}-\frac{\partial G_{\phi r}}{\partial z}-\frac{G_{z\phi}}{r}\right]$$
$$+\hat{\hat{u}}_{r\phi}\left[\frac{1}{r}\frac{\partial G_{z\phi}}{\partial\phi}+\frac{G_{zr}}{r}-\frac{\partial G_{\phi\phi}}{\partial z}\right]+\hat{\hat{u}}_{rz}\left[\frac{1}{r}\frac{\partial G_{zz}}{\partial\phi}-\frac{\partial G_{\phi z}}{\partial z}\right]$$
$$+\hat{\hat{u}}_{\phi r}\left[\frac{\partial G_{rr}}{\partial z}-\frac{\partial G_{zr}}{\partial r}\right]+\hat{\hat{u}}_{\phi\phi}\left[\frac{\partial G_{r\phi}}{\partial z}-\frac{\partial G_{z\phi}}{\partial r}\right]+\hat{\hat{u}}_{\phi z}\left[\frac{\partial G_{rz}}{\partial z}-\frac{\partial G_{zz}}{\partial r}\right]$$
$$+\hat{\hat{u}}_{zr}\left[\frac{1}{r}\frac{\partial\left(rG_{\phi r}\right)}{\partial r}-\frac{1}{r}\frac{\partial G_{rr}}{\partial\phi}+\frac{G_{r\phi}}{r}\right]$$
$$+\hat{\hat{u}}_{z\phi}\left[\frac{1}{r}\frac{\partial\left(rG_{\phi\phi}\right)}{\partial r}-\frac{1}{r}\frac{\partial G_{r\phi}}{\partial\phi}-\frac{G_{rr}}{r}\right]+\hat{\hat{u}}_{zz}\left[\frac{1}{r}\frac{\partial\left(rG_{\phi z}\right)}{\partial r}-\frac{1}{r}\frac{\partial G_{rz}}{\partial\phi}\right]$$
$$\text{(D.2-7)}$$

### D.2.1(c) The gradient of scalar, vector, and dyadic fields

The gradient of a scalar field [from Eq. (4.3-18)] is the vector field

$$\nabla V\Big|_{cyl}=\hat{u}_r\frac{\partial V}{\partial r}+\hat{u}_\phi\frac{1}{r}\frac{\partial V}{\partial\phi}+\hat{u}_z\frac{\partial V}{\partial z} \qquad \text{(D.2-8)}$$

The gradient of a vector field [ref: Eq. (4.3-23)] is the dyadic field

$$\nabla\overline{A}\Big|_{cyl} = +\hat{\hat{u}}_{rr}\frac{\partial A_r}{\partial r} + \hat{\hat{u}}_{r\phi}\frac{\partial A_\phi}{\partial r} + \hat{\hat{u}}_{rz}\frac{\partial A_z}{\partial r}$$
$$+\frac{\hat{\hat{u}}_{\phi r}}{r}\left(\frac{\partial A_r}{\partial \phi} - A_\phi\right) + \frac{\hat{\hat{u}}_{\phi\phi}}{r}\left(\frac{\partial A_\phi}{\partial \phi} + A_r\right) + \frac{\hat{\hat{u}}_{\phi z}}{r}\frac{\partial A_z}{\partial \phi} \qquad \text{(D.2-9)}$$
$$+\hat{\hat{u}}_{zr}\frac{\partial A_r}{\partial z} + \hat{\hat{u}}_{z\phi}\frac{\partial A_\phi}{\partial z} + \hat{\hat{u}}_{zz}\frac{\partial A_z}{\partial z}$$

The gradient of the dyadic is used in the theory of elasticity. See an application of this in Section D.1.1(c). The cylindrical-coordinate expansion of the gradient of a dyadic field is the 27-term triadic field

$$\nabla\overline{\overline{s}}\Big|_{cyl} = \hat{\hat{\hat{u}}}_{rrr}\frac{\partial s_{rr}}{\partial r} + \hat{\hat{\hat{u}}}_{rr\phi}\left[\frac{\partial s_{r\phi}}{\partial r} + \frac{s_{r\phi}}{r}\right] + \hat{\hat{\hat{u}}}_{rrz}\frac{\partial s_{rz}}{\partial r}x$$
$$+\hat{\hat{\hat{u}}}_{r\phi r}\left[\frac{\partial s_{\phi r}}{\partial r} + \frac{s_{\phi r}}{r}\right] + \hat{\hat{\hat{u}}}_{r\phi\phi}\left[\frac{\partial s_{\phi\phi}}{\partial r} + \frac{s_{\phi r}}{r}\right] + \hat{\hat{\hat{u}}}_{r\phi z}\left[\frac{\partial s_{\phi z}}{\partial r} + \frac{s_{\phi z}}{r}\right]$$
$$+\hat{\hat{\hat{u}}}_{rzr}\frac{\partial s_{zr}}{\partial r} + \hat{\hat{\hat{u}}}_{rz\phi}\left[\frac{\partial s_{z\phi}}{\partial r} + \frac{s_{zr}}{r}\right] + \hat{\hat{\hat{u}}}_{rzz}\frac{\partial s_{zz}}{\partial r}$$
$$+\hat{\hat{\hat{u}}}_{\phi rr}\left[\frac{\partial s_{rr}}{\partial \phi} - r\left(s_{\phi r} + s_{r\phi}\right)\right] + \hat{\hat{\hat{u}}}_{\phi r\phi}\left[\frac{\partial s_{r\phi}}{\partial \phi} + \frac{s_{rr}}{r} - rs_{\phi\phi}\right] + \hat{\hat{\hat{u}}}_{\phi rz}\left[\frac{\partial s_{rz}}{\partial \phi} - rs_{\phi z}\right]$$
$$+\hat{\hat{\hat{u}}}_{\phi\phi r}\left[\frac{\partial s_{\phi r}}{\partial \phi} + \frac{s_{rr}}{r} - rs_{\phi\phi}\right] + \hat{\hat{\hat{u}}}_{\phi\phi\phi}\left[\frac{\partial s_{\phi\phi}}{\partial \phi} + \frac{1}{r}\left(s_{r\phi} + s_{\phi r}\right)\right] + \hat{\hat{\hat{u}}}_{\phi\phi z}\left[\frac{\partial s_{\phi z}}{\partial \phi} + \frac{s_{rz}}{r}\right]$$
$$+\hat{\hat{\hat{u}}}_{\phi zr}\left[\frac{\partial s_{zr}}{\partial \phi} - rs_{z\phi}\right] + \hat{\hat{\hat{u}}}_{\phi z\phi}\left[\frac{\partial s_{z\phi}}{\partial \phi} + \frac{s_{zr}}{r}\right] + \hat{\hat{\hat{u}}}_{\phi zz}\frac{\partial s_{zz}}{\partial \phi}$$
$$+\hat{\hat{\hat{u}}}_{zrr}\frac{\partial s_{rr}}{\partial z} + \hat{\hat{\hat{u}}}_{zr\phi}\frac{\partial s_{r\phi}}{\partial z} + \hat{\hat{\hat{u}}}_{zrz}\frac{\partial s_{rz}}{\partial z}$$
$$+\hat{\hat{\hat{u}}}_{z\phi r}\frac{\partial s_{\phi r}}{\partial z} + \hat{\hat{\hat{u}}}_{z\phi\phi}\frac{\partial s_{\phi\phi}}{\partial z} + \hat{\hat{\hat{u}}}_{z\phi z}\frac{\partial s_{\phi z}}{\partial z}$$
$$+\hat{\hat{\hat{u}}}_{zzr}\frac{\partial s_{zr}}{\partial z} + \hat{\hat{\hat{u}}}_{zz\phi}\frac{\partial s_{z\phi}}{\partial z} + \hat{\hat{\hat{u}}}_{zzz}\frac{\partial s_{zz}}{\partial z}$$

(D.2-10)

## D.2.2 Cylindrical coordinate expansions of second-order vector differential operators

### D.2.2(a) The scalar and vector Laplacian

The scalar Laplacian [ref: Eq. (4.7-6)] is the scalar field

$$\nabla^2 V\Big|_{cyl} = \nabla \cdot \nabla V = \frac{1}{r}\frac{\partial}{\partial r}\left(r\frac{\partial V}{\partial r}\right) + \frac{1}{r^2}\frac{\partial^2 V}{\partial \phi^2} + \frac{\partial^2 V}{\partial z^2} \tag{D.2-11}$$

The vector Laplacian [ref: Eq. (4.7-14)] is the vector field

$$\nabla^2 \overline{A}\Big|_{cyl} = \hat{u}_r\left(\nabla^2 A_r - \frac{2}{r^2}\frac{\partial A_\phi}{\partial \phi} - \frac{A_r}{r^2}\right) + \hat{u}_\phi\left(\nabla^2 A_\phi + \frac{2}{r^2}\frac{\partial A_r}{\partial \phi} - \frac{A_\phi}{r^2}\right) + \hat{u}_z \nabla^2 A_z \tag{D.2-12}$$

### D.2.2(b) The curl of the curl of a vector field

The curl of the curl of a vector field [from: Eq. (4.7-23)] is the vector field

$$\begin{aligned}\nabla\times\nabla\times\overline{A}\Big|_{cyl} &= \hat{u}_r\left\{\frac{1}{r^2}\left[\left(\frac{\partial^2(rA_r)}{\partial\phi\partial r} - \frac{\partial^2 A_r}{\partial\phi^2}\right)\right] - \left[\frac{\partial^2 A_r}{\partial z^2} - \frac{\partial^2 A_z}{\partial z\partial r}\right]\right\}\\ &+\hat{u}_\phi\left\{\left[\frac{1}{r}\frac{\partial^2 A_z}{\partial z\partial\phi} - \frac{\partial^2 A_\phi}{\partial z^2}\right] - \left[\frac{1}{r}\frac{\partial^2(rA_\phi)}{\partial r^2} - \frac{1}{r^2}\frac{\partial(rA_\phi)}{\partial r} + \frac{1}{r}\frac{\partial^2 A_r}{\partial x\partial\phi} - \frac{1}{r^2}\frac{\partial A_r}{\partial\phi}\right]\right\}\\ &+\hat{u}_z\left\{\frac{\partial^2 A_r}{\partial r\partial z} + \frac{1}{r}\frac{\partial A_r}{\partial z} - \frac{\partial^2 A_z}{\partial r^2} - \frac{1}{r}\frac{\partial A_z}{\partial r} - \frac{1}{r^2}\frac{\partial^2 A_z}{\partial\phi^2} + \frac{1}{r}\frac{\partial^2 A_\phi}{\partial\phi\partial z}\right\}\end{aligned} \tag{D.2-13}$$

### D.2.2(c) The gradient of the divergence

The gradient of the divergence of a vector field in cylindrical coordinates from Eq. (4.7-24) is the vector field

$$\nabla\nabla\cdot\bar{A}=\hat{u}_r\frac{1}{r^2}\left[r\frac{\partial^2(rA_r)}{\partial r^2}-\frac{\partial(rA_r)}{\partial r}+r\frac{\partial^2 A_\phi}{\partial r\partial\phi}-\frac{\partial A_\phi}{\partial\phi}+r^2\frac{\partial^2 A_z}{\partial r\partial z}\right]$$

$$+\hat{u}_\phi\frac{1}{r^2}\left[\frac{\partial^2(rA_r)}{\partial\phi\partial r}+\frac{\partial^2 A_\phi}{\partial\phi^2}+r\frac{\partial^2 A_z}{\partial\phi\partial z}\right]+\hat{u}_z\left[\frac{1}{r}\frac{\partial^2(rA_r)}{\partial z\partial r}+\frac{1}{r}\frac{\partial^2 A_\phi}{\partial z\partial\phi}+\frac{\partial^2 A_z}{\partial z^2}\right]$$

(D.2-14)

See the inside back cover as a quick reference for some of the more common first- and second-order vector differential operator expansions in cylindrical coordinates. The inside front cover displays the same selected common vector operator expansions in Cartesian Coordinates.

## Reference

1. A. I, Lure, *Three-Diminsional Problems of the Theory of Elasticity*, D. B. MacVean, Trans., Wiley Interscience, Hoboken, NJ (1964), p. 53.

# Glossary

**Ampere's circuital law:** The circulation of the magnetic field intensity $\bar{H}$ about any closed path is equal to the total electric current passing through any surface bounded by that closed path.

**curl of a vector field at a point:** A vector pointing in the direction of a normal to an infinitesimal surface which is so oriented in space that the limit of the ratio of the line integral of the vector field around the perimeter of that surface to the area enclosed is *maximal*. The magnitude of the curl is the value of that limit.

**curl of the curl of a vector field:** The circulation density of the vorticity of that field, which can be thought of as the rotational spatial change of vorticity in the cross-product direction.

**divergence of a vector field:** A scalar field whose magnitude at any point in space is determined by taking the ratio of the net outward flux of the vector field through an infinitesimal closed surface surrounding the point to the volume enclosed by that surface as the volume tends to zero.

**directional compoundedness:** An integer that denotes the level of directionality of a field quantity. The tensor rank of that quantity. A term coined by the author intended to give those new to tensor fields a more intuitive feel for tensor rank.

**dyad:** See unit dyad.

**dyadic:** A quantity that has two directions associated with each point in the field. A tensor of rank two.

**dyadic field:** A field quantity that has a dual directional compoundedness. A quantity that has two directions associated with each point in space. A tensor field of rank two.

**field:** A quantity that is a function of spatial coordinates.

**Gauss' law for electrostatics:** The total electric flux emanating outwardly through a closed surface is equal to the total charge enclosed within.

**GOCCs:** generalized orthogonal curvilinear coordinates.

**gradient of a scalar field:** A vector field oriented in the direction in which the scalar field increases most rapidly. Its magnitude is the derivative of the scalar field in the direction of its maximal increase.

**gradient of a vector field:** A dyadic field found by taking the nabla differential operation on each of the vector field components.

**gradient of the divergence of a vector field:** Another vector field oriented in the direction in which the volume source distribution density increases most rapidly. Its magnitude is the derivative of that volume density distribution in the direction of its maximal increase.

**line integral:** The integral of a field quantity taken over a vector differential length $\overline{d\ell}$ that is everywhere tangent to a general line $L$ in space.

**mixed derivative theorem:** States that if $f(q_1,q_2,q_3)$ and its partial derivatives $f_{q_1}$, $f_{q_1}$, and $f_{q_1q_2}$ exist and are continuous, then $f_{q_2q_1}$ also exists and $f_{q_1q_2} = f_{q_2q_1}$.

**order of a tensor:** In modern photonics (and in this book) the order of a tensor field is one less than the rank. In other treatments order is sometimes used synonymously with rank.

**partial derivative:** The result of taking a derivative of a function of multiple independent variables with respect to one of the variables while holding all of the other independent variables constant.

**quadad:** See unit quadadic.

**quadadic:** A quantity that has four directions associated with each point in the field. A tensor of rank four.

**quadadic field:** A field quantity that has quadruple directional compoundedness. A quantity that has four directions associated with each point in space. A tensor field of rank four.

**$\overline{r}$-space notation:** A vector-like symbol $\overline{r}$ used in the functional argument of a field quantity to denote the coordinates at which function is being represented. A shorthand notation for those coordinates, e. g., $\overline{r}$ is short for $x,y,z$.

**rank:** The quantitative (integer) property of a tensor that specifies its directional compoundedness or the multiplicity of its directionality.

**scalar:** A quantity that has a magnitude but no directionality. A tensor of rank zero.

**surface integral:** the integral of a field quantity taken over a vector differential area $\overline{da}$ that is everywhere normal to a general surface S in space.

**tensor:** A quantity that has multiple directionality at each point in space and at each moment in time. The "rank" of a tensor enumerates that multiplicity.

**tensor field:** A quantity that has multiple directionality at each point in space. A quantity with arbitrary (integer) directional compoundedness. The "rank" of a tensor enumerates that multiplicity.

**triad:** See unit triad.

**triadic:** A quantity that has three directions associated with each point in the field. A tensor of rank three.

**triadic field:** A field quantity that has a triple directional compoundedness. A quantity that has three directions associated with each point in the field. A tensor field of rank three.

**unit dyad:** A dually directed unitary dyadic. A quantity that has a magnitude of one, and two directions at every point in space. A unitary tensor of rank two.

**unit quadad:** A quadruply directed unitary quadadic. A quantity that has a magnitude of one, and four directions at every point in space. A unitary tensor of rank four.

**unit triad:** A triply directed unitary triadic. A quantity that has a magnitude of one, and three directions at every point in space. A unitary tensor of rank three.

**unit vector:** A singly directed unitary vector. A quantity that has a magnitude of one, and a single direction at every point in space. A unitary tensor of rank one.

**vector:** A quantity that has a magnitude and an inherent *single* direction. A tensor of rank one.

**vector field:** A quantity that has a magnitude and an inherent single direction at every point in the field. A tensor field of rank one.

# Index

Bernard Maxum received his Ph.D. in Electrical Engineering from the University of California, Berkeley following his MSEE from the University of Southern California and his BSEE from the University of Washington, Seattle. He worked for two decades in aerospace industrial R&D. In 1992 he joined Lamar University, a member of the Texas State University System, where he has served for nine years as Chair of the Department of Electrical Engineering and is now Professor of Optical Communications.